The Origins of Modern Science 1300–1800

HERBERT BUTTERFIELD

The Origins of Modern Science 1300-1800

Revised Edition

Fp

THE FREE PRESS, *New York*

Printed in the United States of America

FIRST FREE PRESS PAPERBACK EDITION 1965

printing number
10

Contents

Introduction

CONSIDERING THE PART played by the sciences in the story of our Western civilisation, it is hardly possible to doubt the importance which the history of science will sooner or later acquire both in its own right and as the bridge which has been so long needed between the Arts and the Sciences.

The following lectures, which were delivered for the History of Science Committee in Cambridge in 1948, were produced in the hope that they would stimulate in the historian a little interest in science, and in the scientist a little interest in history. In this revised edition they appear now with some of their original errors removed, some judgments altered, and some changes which reflect the advance of knowledge in the intervening years.

Nobody, of course, will imagine that the mere "general historian" can pretend to broach the question of the more recent developments in any of the natural sciences; but it is fortunate that in respect of students in both the Arts and the Sciences the supremely important field for the ordinary purposes of education is one more manageable in itself and, indeed, perhaps more in need of the intervention of the historian as such. It is the so-called "scientific revolution," popularly associated with the sixteenth and seventeenth centuries, but reaching back in an unmistakably continuous line to a period much earlier still. Since that revolution overturned the authority in science not only of the middle ages but of the ancient world—since it ended not only in the eclipse of scholastic philosophy but in the destruction of Aristotelian physics—it outshines everything since the rise of Christianity and reduces the Renaissance and Reformation to the rank of mere episodes, mere internal displacements, within the system of medieval Christendom. Since it changed the character of men's habitual mental operations even in the conduct of the non-material sciences, while transforming the whole diagram of the physical universe and the very texture of human life itself, it looms so large as the real origin both of the modern world and of the modern mentality that our customary periodisation of European history has become an anachronism and an encumbrance. There can hardly be a field in which it is of greater moment to us to see at somewhat closer range the precise operations that underlay a particular historical transition, a particular chapter of intellectual development.

It is this phase of European history which the following course of lectures is chiefly intended to survey. There will be no pretence of laying out four centuries of the history of science like a long piece of wall-paper, however, and dividing it into so many units of superficial area—so much acreage of historical narrative to be covered in each lecture after the manner of the encyclopaedist and the abridger. It will be necessary rather to look for the lines of strategic change, and to put the microscope upon those moments that seem pivotal—trying, for example, to discover the particular intellectual knots that had to be untied at a given conjuncture. It will concern us particularly to take note of those cases in which men not only solved a problem but had to alter their mentality in the process, or at least discovered afterwards that the solution involved a change in their mental approach.

Two further points of method ought perhaps to be noted in order to avoid misunderstanding and cross-purposes. First of all, the subject has not been turned into genuine history—it is still at an inferior degree of organisation, like the work of the annalist and the chronicler—if we remain content with a merely biographical mode of treatment, and particularly if we construct our story of science by drawing lines straight from one great figure to another. Some of the surprises, some of the remarkable reversals of judgment, that have taken place in the last fifty years have been the result of a much more detailed study of a host of intervening scientific workers whose names had been comparatively unknown. Secondly, the whole fabric of our history of science is lifeless and its whole shape is distorted if we seize upon this particular man in the fifteenth century who had an idea that strikes us as modern, now upon another man of the sixteenth century who had a hunch or an anticipation of some later theory—all as if one were making a catalogue of inventions or of maritime discoveries. It has proved almost more useful to learn something of the misfires and the mistaken hypotheses of early scientists, to examine the particular intellectual hurdles that seemed insurmountable at given periods, and even to pursue courses of scientific development which ran into a blind alley, but which still had their effect on the progress of science in general. Similarly in these lectures we may try to examine various facets or aspects of what is called the scientific revolution; we shall not be able to measure the achievement at any given moment, however, if we merely pay attention to the new doctrines and take note of the emergence of the views that we now regard as right. It is necessary on each occasion to have a picture of the older systems—the type of science that was to be overthrown.

Finally, it is relevant to note that, in a still larger sense, we must proceed in the history of science from the earlier to the latter—from the sixteenth-century ideas of mechanics to the ideas of Galileo—so that we can know exactly how a great thinker operated on the margin of contemporary thought, or created a new synthesis, or completed a line of development already taking place. It is not sufficient to read Galileo with the eyes of the twentieth century or to interpret him in modern terms—we can only understand his work if we know something of the system which he was attacking, and we must know something of that system apart from the things which were said about it by its enemies. In any case, it is necessary not merely to describe and expound discoveries, but to probe more deeply into historical processes and to learn something of the interconnectedness of events, as well as to exert all our endeavours for the understanding of men who were not like-minded with ourselves. Little progress can be made if we think of the older studies as merely a case of bad science or if we imagine that only the achievements of the scientist in very recent times are worthy of serious atten-

The Origins of Modern Science 1300–1800

Chapter 1

The Historical Importance of a Theory of Impetus

IT IS ONE of the paradoxes of the whole story with which we have to deal that the most sensational step leading to the scientific revolution in astronomy was taken long before the discovery of the telescope—even long before the Danish astronomer, Tycho Brahe, in the latter part of the sixteenth century, had shown the great improvement that it was still possible to achieve in observations made with the naked eye. When William Harvey in England opened up new paths for physiology by his study of the action of the heart, he alluded once or twice to his use of a magnifying glass, but he carried out his revolutionary work before any serviceable kind of microscope had become available. With regard to the transformation of the science of mechanics, it is remarkable to what an extent even Galileo discusses the ordinary phenomena of everyday life, conjectures what would happen if a stone were thrown from the mast of a moving ship, or plays with pellets on inclined planes in a manner that had long been customary. In fact, we shall find that in both celestial and terrestrial physics—which hold the strategic place in the whole movement—change is brought about, not by new observations or additional evidence in the first instance, but by transpositions that were taking place inside the minds of the scientists themselves. In this connection it is not irrelevant to note that, of all forms of mental activity, the most difficult to induce even in the minds of the young, who may be presumed not to have lost their flexibility, is the art of handling the same bundle of data as before, but placing them in a new system of relations with one another by giving them a different framework, all of which virtually means putting on a different kind of thinking-cap for the moment. It is easy to teach anybody a new fact about Richelieu, but it needs light from heaven to enable a teacher to break the old framework in which the student has been accustomed to seeing his Richelieu—the framework which is built up sometimes far too rigidly by the Higher Certificate student, and into which he will fit whatever new information he ever afterwards acquires on this subject. But the supreme paradox of the scientific revolution is the fact that things which we find it easy to instil into boys at school, because we see that they start off on the right foot—things which would strike us as the ordinary natural way of looking at the universe, the obvious way of regarding the behaviour of falling bodies, for example—defeated the greatest intellects for centuries, defeated Leonardo da Vinci and at the marginal point even Galileo, when their minds were wrestling on the very frontiers of human thought with these very problems. Even the greatest geniuses who broke through the ancient views in some special field of study—Gilbert, Bacon and Harvey, for example—would remain stranded in a species of medievalism when they went outside that chosen field. It required their combined efforts to clear up certain simple things which we should now regard as obvious to any unprejudiced mind, and even easy for a child.

A particular development of ideas which was already taking place in the later middle ages has come to stand as the first chapter in the history of the transition to what we call the scientific revolution. It is a field of thought upon which an expositor can embark only with the greatest trepidation, in view of the vicissitudes of lecturers at the very beginning of modern times. Students of history will remember how the humanists of the Renaissance, Erasmus included, were accustomed to complaining of the boredom—deriding the sophistries and subtleties—of the scholastic lectures which they had to endure at the university. Occasionally they specified the forms of teaching and lecturing to which they most objected, and as they particularly mentioned those discussions of mechanics with which we have now to concern ourselves, it will no doubt be prudent to make the examination of such teaching as brief as possible. It is curious that these despised scholastic disquisitions should now have come to hold a remarkable key-position in the story of the evolution of the modern mind. Perhaps the lack of mathematics, or the failure to think of mathematical ways of formulating things, was partly responsible for what appeared to be verbal subtleties and an excessive straining of language in these men who were almost yearning to find the way to the modern science of mechanics.

Of all the intellectual hurdles which the human mind has confronted and has overcome in the last fifteen hundred years, the one which seems to me to have been the most amazing in character and the most stupendous in the scope of its consequences is the one relating to the problem of motion—the one which perhaps was hardly disposed of by Galileo, though it received a definitive form of settlement shortly after his time in the full revised statement of what every schoolboy learns to call the law of inertia. On this question of motion the Aristotelian teaching, precisely because it carried such an intricate dovetailing of observations and explanations—that is to say, precisely because it was part of a system which was such a colossal intellectual feat in itself—was hard for the human mind to escape from, and gained a strong hold on medieval scholastic thought. Furthermore, it remains as the essential background of the story—it continues to present the presiding issue—until the time of Galileo himself; in other words, until the first half of the seventeenth century. On the Aristotelian theory all heavy terrestrial bodies had a natural motion towards the centre of the universe, which for medieval thinkers was at or near the centre of the earth; but motion in any other direction was violent motion, because it contradicted the ordinary tendency of a body to move to what was regarded as its natural place. Such motion depended on the operation of a mover, and the Aristotelian doctrine of inertia was a doctrine of rest—it was motion, not rest, that always required to be explained. Wherever this motion existed, and however long it existed, something had to be brought in to account for it.

The essential feature of this view was the assertion or the assumption that a body would keep in movement only so long as a mover was actually in contact with it, imparting motion to it all the time. Once the mover ceased to operate, the movement stopped—the body fell straight to earth or dropped suddenly to rest. Further—a point that will seem very heretical to the present day—it was argued that, provided the resistance of the medium through which the body passed remained a constant, the speed of the body would be proportionate to what we should describe as the force consistently being exerted upon it by the mover. A constant force exerted by the mover over a given length of time produced not any acceleration at all, but a uniform motion for the whole period. On the other hand, if there was any variation in the resistance of the medium—the difference between moving in air and moving in water, for example—the speed would vary in inverse proportion to this, provided the other factors remained constant. And if the resistance were reduced to nought, the speed would be infinite; that is to say, if the movement took place in a vacuum, bodies would move from one place to another instantaneously. The absurdity of this was one of the reasons why the Aristotelians regarded a complete void as impossible, and said that God Himself could not make one.

It is astonishing to what a degree not only this theory but its rivals—even the ones which superseded it in the course of the scientific revolution—were based on the ordinary observation of the data available to common sense. And, as writers have clearly pointed out, it is not relevant for us to argue that if the Aristotelians had merely watched the more carefully they would have changed their theory of inertia for the modern one—changed over to the view that bodies tend to continue either at rest or in motion along a straight line until something intervenes to stop them or deflect their course. It was supremely difficult to escape from the Aristotelian doctrine by merely observing things more closely, especially if you had already started off on the wrong foot and were hampered beforehand with the whole system of interlocking Aristotelian ideas. In fact, the modern law of inertia is not the thing you would discover by mere photographic methods of observation—it required a different kind of thinking-cap, a transposition in the mind of the scientist himself; for we do not actually see ordinary objects continuing their rectilinear motion in that kind of empty space which Aristotle said could not occur, and sailing away to that infinity which also he said could not possibly exist; and we do not in real life have perfectly spherical balls moving on perfectly smooth horizontal planes—the trick lay in the fact that it occurred to Galileo to imagine these. Furthermore, even when men were coming extraordinarily near to what we should call the truth about local motion, they did not clinch the matter—the thing did not come out clear and clean—until they had realised and had made completely conscious to themselves the fact that they were in reality transposing the questions into a different realm. They were discussing not real bodies as we actually observe them in the real world but geometrical bodies moving in a world without resistance and without gravity—moving in that boundless emptiness of Euclidean space which Aristotle had regarded as unthinkable. In the long run, therefore, we have to recognise that here was a problem of a fundamental nature, and it could not be solved by close observation within the framework of the older system of ideas—it required a transposition in the mind.

As often happened with such theories in those days, if not now, the Aristotelian doctrine of motion might seem to correspond in a self-evident manner with most of the data available to common sense, but there were small pockets of fact which did not square with the theory at the first stage of the argument; they were unamenable to the Aristotelian laws at what we should call the ordinary common-sense level. There were one or two anomalies which required a further degree of analysis before they could be satisfactorily adjusted to the system; and perhaps, as some writers have said, the Aristotelian theory came to a brilliant peak in the manner by which it hauled these exceptional cases into the synthesis and established (at a second remove) their conformity with the stated rules. On the argument so far as we have taken it, an arrow ought to have fallen to the ground the moment it lost contact with the bow-string; for neither the bow-string nor anything else could impart a motion which would continue after the direct contact with the original mover had been broken. The Aristotelians explained the continued movement of projectiles by the commotion which the initial movement had produced in the air—especially as the air which was being pushed and compressed in front had to rush round behind to prevent that vacuum which must never be allowed to take place. At this point in the argument there even occurred a serious fault in observation which harassed the writers on physical science for many centuries. It was thought that the rush of air produced an actual initial acceleration in the arrow after it had left the bow-string, and it is curious to note that Leonardo da Vinci and later writers shared this mistake—the artillerymen of the Renaissance were victims of the same error—though there had been people in the later middle ages who had taken care not to commit themselves on this point. The motion of a projectile, since it was caused by a disturbance in the medium itself, was a thing which it was not possible to imagine taking place in a vacuum.

Furthermore, since the Aristotelian commentators held something corresponding to the view that a constant uniform force only produced uniform motion, there was a second serious anomaly to be explained—it was necessary to produce special reasons to account for the fact that falling bodies were observed to move at an accelerating speed. Once again the supporters of the older teaching used the argument from the rush of air, or they thought that, as the body approached the earth, the greater height of the atmosphere above meant an increase in the downward pressure, while the shorter column of air below would offer a diminishing resistance to the descent. Alternatively they used Aristotle's argument that the falling body moved more jubilantly every moment because it found itself nearer home.

From the fourteenth to the seventeenth century, then, this Aristotelian doctrine of motion persisted in the face of recurrent controversy, and it was only in the later stages of that period that the satisfactory alternative emerged, somewhat on the policy of picking up the opposite end of the stick. Once this question was solved in the modern manner, it altered much of one's ordinary thinking about the world and opened the way for a flood of further discoveries and reinterpretations, even in the realm of common sense, before any very elaborate experiments had been embarked upon. It was as though science or human thought had been held up by a barrier until this moment—the waters dammed because of an initial defect in one's attitude to everything in the universe that had any sort of motion—and now the floods were released. Change and discovery were bound to come in cascades even if there were no other factors working for a scientific revolution. Indeed, we might say that a change in one's attitude to the movement of things that move was bound to result in so many new analyses of various kinds of motion that it constituted a scientific revolution in itself.

Apart from all this there was one special feature of the problem which made the issue momentous. We have not always brought home to ourselves the peculiar character of that Aristotelian universe in which the things that were in motion had to be accompanied by a mover all the time. A universe constructed on the mechanics of Aristotle had the door half-way open for spirits already; it was a universe in which unseen hands had to be in constant operation, and sublime Intelligences had to roll the planetary spheres around. Alternatively, bodies had to be endowed with souls and aspirations, with a "disposition" to certain kinds of motions, so that matter itself seemed to possess mystical qualities. The modern law of inertia, the modern theory of motion, is the great factor which in the seventeenth century helped to drive the spirits out of the world and opened the way to a universe that ran like a piece of clockwork. Not only so—but the very first men who in the middle ages launched the great attack on the Aristotelian theory were conscious of the fact that this colossal issue was involved in the question. One of the early important figures, Jean Buridan in the middle of the fourteenth century, pointed out that his alternative interpretation would eliminate the need for the Intelligences that turned the celestial spheres. He even noted that the Bible provided no authority for these spiritual agencies—they were demanded by the teaching of the ancient Greeks, not by the Christian religion as such. Not much later than this, Nicholas of Oresme went further still, and said that, on the new alternative theory, God might have started off the universe as a kind of clock and left it to run of itself.

Ever since the earlier years of the twentieth century at latest, therefore, a great and growing interest has been taken in that school of thinkers who so far back as the fourteenth century were challenging the Aristotelian explanations of motion, and who put forward an alternative doctrine of "impetus" which—though imperfect in itself—must represent the first stage in the history of the scientific revolution. And if it is imagined that this kind of argument falls into one of the traps which it is always necessary to guard against—picking out from the middle ages mere anticipations and casual analogies to modern ideas—the answer to that objection will be clear to us if we bear in mind the kind of rules that ought to govern historians in such matters. Here we have a case of a consistent body of teaching which rises in Oxford, is developed as a tradition by a school of thinkers in Paris, and is still being taught in Paris at the beginning of the sixteenth century. It has a continuous history—we know how this teaching passed into Italy, how it was promulgated in the universities of the Renaissance, and how Leonardo da Vinci picked it up, so that some of what were once considered to be remarkable strokes of modernity, remarkable flashes of genius, in his notebooks, were in reality transcriptions from fourteenth-century Parisian scholastic writers. We know how the teaching was developed in Italy later in the sixteenth century, how it was misunderstood on occasion—sometimes only partially appropriated—and how some of Galileo's early writings on motion are reminiscent of this school, being associated with that doctrine of the "impetus" which it is our purpose to examine. It is even known fairly certainly in what edition Galileo read the works of certain writers belonging to this fourteenth-century Parisian school. Indeed, Galileo could have produced much, though not quite all, that we find in his juvenile works on this particular subject if he had lived in the fourteenth century; and in this field one might very well ask what the world with its Renaissance and so forth had been doing in the meantime. It has been suggested that if printing had been invented two centuries earlier the doctrine of "impetus" would have produced a more rapid general development in the history of science, and would not have needed so long to pass from the stage of Jean Buridan to the stage of Galileo.

If the orthodox doctrine of the middle ages had been based on Aristotle, however, it has to be noted that, both then and during the Renaissance (as well as later still), the attacks on Aristotle—the theory of impetus included—would themselves be based on some ancient thinker. Here we touch on one of the generative factors, not only in the formation of the modern world, but also in the development of the scientific revolution—namely, the discovery of the fact that even Aristotle had not reigned unchallenged in the ancient days. All this produced a healthy friction, resulting in the emergence of important problems which the middle ages had to make up their own minds about, so that men were driven to some kind of examination of the workings of nature themselves, even if only because they had to decide between Aristotle and some rival teacher. It also appears that a religious factor affected the rise of that movement which produced the theory of impetus, and, in a curious manner which one tries in vain to analyse away, a religious taboo operated for once in favour of freedom for scientific hypothesis. In the year 1277 a council in Paris condemned a large number of Aristotelian theses, such as the view that even God could not create a void, or an infinite universe, or a plurality of worlds; and that decision—one in which certain forms of partisanship were involved—was apparently extended by the Archbishop of Canterbury to this country. The regions that came within the orbit of these decisions must have been the seat of a certain anti-Aristotelian bias already; and certainly from this time both Oxford and Paris showed the effects of this bias in the field of what we should call physical science. From this time also the discussion of the possibility of the existence of empty space, or of an infinite universe, or of a plurality of worlds takes a remarkable step forward in Paris. And amongst the names concerned in this development are some which figure in the rise of the doctrine of impetus. It has been pointed out furthermore that in the same Parisian tradition there was a tendency towards something in the nature of mathematical physics, though the mathematics of the time were not sufficiently advanced to allow of this being carried very far or to produce anything like the achievement of Galileo in the way of a mathematical approach to scientific problems. We must avoid the temptation, however, to stress unduly the apparent analogies with modern times and the "anticipations" which are so easy to discover in the past—things which often owe a little, no doubt, to the trick-mirrors of the historian. And though it may be useful sometimes, in order to illustrate a point, we must beware of submitting to the fascination of "what might have been."

The people who chiefly concern us, then, are certain fourteenth-century writers, first of all a group at Merton College, Oxford, and, after these, Jean Buridan, Albert of Saxony and Nicholas of Oresme. They are important for other things besides their teaching on the subject of impetus. The contemporaries of Erasmus laughed at the scholastic lecturers for discussing not only "uniform motion" and "difform motion," but also "uniform difform motion"—all carried to a great degree of subtlety—but it transpired in the sixteenth century, when the world was looking for a formula to represent the uniform acceleration of falling bodies, that the solution of the problem had been at their disposal for a long time in the medieval formula for the case of uniformly difform motion. The whole development which we are studying took place amongst people who, in fact, were working upon questions and answers which had been suggested by Aristotle. These people came up against the Aristotelian theory of motion at the very points where we should expect the attack to take place, namely, in connection with the two particularly doubtful questions of the movement of projectiles and the acceleration of falling bodies. If we take a glance at the kind of arguments they used, we can observe the type of critical procedure which would take place even in the middle ages, producing changes on the margin of the current Aristotelian teaching. We are observing also the early stages of the great debate on certain issues that lay at the heart of the scientific revolution itself. Indeed, the arguments which were employed at this early period often reappeared —with reference to precisely the same instances—even in the major works of Galileo, for they passed into general currency in the course of time. And if they seem simple arguments based on the ordinary phenomena available to common sense, we ought to remember that many of the newer arguments brought forward by Galileo himself at a later stage of the story were really of the same type.

According to the view developed by these thinkers, the projectile was carried forward by an actual impetus which it had acquired and which bodies were capable of acquiring, from the mere fact of being in motion. And this impetus was supposed to be a thing inside the body itself—occasionally it was described as an impetuosity that had been imparted to it; occasionally one sees it discussed as though it were itself movement which the body acquired as the result of being in motion. In any case this view made it possible for men to contemplate the continued motion of a body after the contact with the original mover had been lost. It was explained that the impetus lay in the body and continued there, as heat stays in a red-hot poker after it has been taken from the fire; while in the case of falling bodies the effect was described as accidental gravity, an additional gravity which the body acquired as a result of being in motion, so that the acceleration of falling bodies was due to the effects of impetus being continually added to the constant fall due to ordinary weight. A constant force exerted on a body, therefore, produced here not uniform motion but a uniform rate of acceleration. It is to be noted, however, that Leonardo da Vinci, like a number of others who accepted the general theory of impetus, failed to follow the Parisian school in the application of their teaching to the acceleration of falling bodies. Whereas the Aristotelians thought that falling bodies rushed more quickly as they got nearer home, the new teaching inverted this, and said that it was rather the distance from the starting-point that mattered. If two bodies fell to earth along the same line BC, the one which had started higher up at A would move more quickly from B to C than the one that started at B, though in this particular part of their course they were both equally distant from the centre of the earth. It followed from the new doctrine that if a cylindrical hole were cut through the earth, passing through the centre, a body, when it reached the centre, would be carried forward on its own impetus for some distance, and indeed would oscillate about the centre for some time—a thing impossible to conceive under the terms of the ancient theory. There was a further point in regard to which the Aristotelians had been unconvincing; for if the continued flight of a projectile were really due not to the thrower but to the rush of air, it was difficult to see why the air should carry a stone so much farther than a ball of feathers—why one should be able to throw the stone the greater distance. The newer school showed that, starting at a given pace, a greater impetus would be communicated to the stone by reason of the density of its material than to a feather; though, of course, a larger body of the same material would not travel farther—a large stone would not be more easy to throw than a small one. Mass was used as the measure of the impetus which corresponded with a given speed.

Since Aristotle found it necessary on occasion to regard the air as a resisting factor, he was open to the charge that one could not then—in the next breath, so to speak—start using the argument that the air was also the actual propellant. The new school said that the air could not be the propellant except in the case of a high wind; and they brought the further objection that if the original perturbation of the air—the rush which occurred when the bow-string started the arrow—had the capacity to repeat itself, pushing the arrow on and on, there could be no reason why it should ever stop; it ought to go on for ever repeating itself, communicating further perturbations to every next region of the atmosphere. Furthermore, a thread tied to a projectile ought to be blown ahead of it, instead of trailing behind. In any case, on the Aristotelian view of the matter it ought to be impossible for an arrow to fly against the wind. Even the apostles of the new theory of impetus, however, regarded a projectile as moving in a straight line until the impetus had exhausted itself, and then quickly curving round to make a direct vertical drop to earth. They looked upon this impetus as a thing which gradually weakened and wore itself out, just as a poker grows cold when taken from the fire. Or, said Galileo, it was like the reverberations which go on in a bell long after it has been struck, but which gradually fade away. Only, in the case of the celestial bodies and the orbs which carried the planets round the sky, the impulse never exhausted itself—the pace of these bodies never slackened since there was no air-resistance to slow them down. Therefore, it could be argued, God might have given these things their initial impetus, and their motion could be imagined as continuing for ever.

The theory of the impetus did not solve all problems, however, and proved to be only the half-way house to the modern view, which is fairly explicit in Galileo though it received its perfect formulation only in Descartes—the view that a body continues its motion in a straight line until something intervenes to halt or slacken or deflect it. As I have already mentioned, this modern law of inertia is calculated to present itself more easily to the mind when a transposition has taken place—when we see, not real bodies, moving under the restrictions of the real world and clogged by the atmosphere, but geometrical bodies sailing away in empty Euclidean space. Archimedes, whose works were more completely discovered at the Renaissance and became very influential especially after the translation published in 1543, appears to have done something to assist and encourage this habit of mind; and nothing could have been more important than the growing tendency to geometrise or mathematise a problem. Nothing is more effective, after people have long been debating and wrangling and churning the air, than the appearance of a person who draws a line on the blackboard, which with the help of a little geometry solves the whole problem in an instant. In any case, it is possible that Archimedes, who taught people to think of the weight of a thing in water and then its weight in air and finally, therefore, its weight when unencumbered by either, helped to induce some men to pick up the problem of motion from the opposite end to the usual one, and to think of the simplest form of motion as occurring when there was no resisting medium to complicate it. So you assumed the tendency in bodies to continue their existing motion along a straight line, and you set about afterwards to examine the things which might clog or hamper or qualify that motion; whereas Aristotle, assuming that the state of rest was natural and that bodies tended to return to it when left to themselves, had the difficult task of providing an active mover that should operate as long as the body continued to have any movement at all.

On the other hand, it may be true to say that Aristotle, when he thought of motion, had in mind a horse drawing a cart, so that his whole feeling for the problem was spoiled by his preoccupation with a misleading example. The very fact that his teaching on the subject of projectiles was so unsatisfactory may have helped to produce the phenomenon of a later age which, when it thought of motion, had rather the motion of projectiles in mind, and so acquired a different feeling in regard to the whole matter.

It is natural that the transition to modern science should often appear to us as a reaction against the doctrines of Aristotle. Because there was a conservative resistance to be combated, the supporters of the new ideas would feel compelled to produce what was sometimes a bitterly anti-Aristotelian polemical literature. Appearances are deceptive, however, and often it is fairer to regard the new ideas as the developing achievement of the successive commentators on Aristotle. These men realised their indebtedness to the ancient master; and they would hold to a great part of his system even if, at one place and another, they were pressing against the frontiers of that system. In answer to the conservatives of their time, the innovators would sometimes argue that Aristotle himself would have been on their side if he had been living in the modern world. The conflicts of the later medieval and early modern centuries ought not to be allowed to diminish our impression of the greatness of this ancient teacher, who provoked so much thought and controversy, and who kept the presiding position for so long. Nor ought we to imagine that Aristotle shared the faults of those people who, in the sixteenth and seventeenth centuries, would be held to be of the "Aristotelian" party merely because they were conservatives.

The work of Pierre Duhem, who, over fifty years ago, brought out the importance of the fourteenth-century teaching on the subject of the "impetus," has not remained free from criticism in the period that has since elapsed. On the one hand the story has been carried behind Jean Buridan and the Parisian school—carried further back to Merton College, Oxford. On the other hand it has been rightly pointed out that the transition from the doctrine of the "impetus" to the modern doctrine of inertia required—from Galileo, for example—greater originality than some writers seem to allow. It is also true that the originality in the fourteenth-century writers extended beyond the problem of motion which we have been considering; and by this time, as we shall see, advance was already taking place in the theoretical discussion of scientific method. It is possible to exaggerate the rôle of these medieval precursors, and so to under-estimate the magnitude of the seventeenth-century revolution. But the work of Duhem in the field that we have been considering has been an important factor in the great change which has taken place in the attitude of historians of science to the middle ages. One of the strands of the historical narrative with which we are concerned is the progress which is made on occasion through the development of scholastic thinking itself.

Chapter 2

The Conservatism of Copernicus

AN INTRODUCTORY sketch of the medieval view of the cosmos must be qualified first of all with the reservation that in this particular realm of thought there were variations, uncertainties, controversies and developments which it would obviously be impossible to describe in detail. On the whole, therefore, it would be well, perhaps, if we were to take Dante's view of the universe as a pattern, because it will be easy to note in parenthesis some of the important variations that occurred, and at the same time this policy will enable us to see in a single survey the range of the multiple objections which it took the Copernican theory something like a hundred and fifty years to surmount.

According to Dante, what one must have in mind is a series of spheres, one inside another, and at the heart of the whole system lies the motionless earth. The realm of what we should call ordinary matter is confined to the earth and its neighbourhood—the region below the moon; and this matter, the stuff that we can hold between our fingers and which our modern physical sciences set out to study, is humble and unstable, being subject to change and decay for reasons which we shall examine later. The skies and the heavenly bodies—the rotating spheres and the stars and planets that are attached to them—are made of a very tangible kind of matter too, though it is more subtle in quality and it is not subject to change and corruption. It is not subject to the physical laws that govern the more earthy kind of material which we have below the moon. From the point of view of what we should call purely physical science, the earth and the skies therefore were cut off from one another and, for a medieval student, were separate organisations, though in a wider system of thought they dovetailed together to form one coherent cosmos.

As to the ordinary matter of which the earth is composed, it is formed of four elements, and these are graded according to their virtue, their nobility. There is earth, which is the meanest stuff of all, then water, then air and, finally, fire, and this last comes highest in the hierarchy. We do not see these elements in their pure and undiluted form, however—the earthy stuff that we handle when we pick up a little soil is a base compound and the fire that we actually see is mixed with earthiness. Of the four elements, earth and water possess gravity; they have a tendency to fall; they can only be at rest at the centre of the universe. Fire and air do not have gravity, but possess the very reverse; they are characterised by levity, an actual tendency to rise, though the atmosphere clings a little to the earth because it is loaded with base mundane impurities. For all the elements have their spheres, and aspire to reach their proper sphere, where they find stability and rest; and when flame, for example, has soared to its own upper region it will be happy and contented, for here it can be still and can most endure. If the elements did not mix—if they were all at home in their proper spheres—we should have a solid sphere of earth at the heart of everything and every particle of it would be still. We should then have an ocean covering that whole globe, like a cap that fitted all round, then a sphere of air, which far above mountaintops was supposed to swirl round from east to west in sympathy with the movement of the skies. Finally, there would come the region of enduring fire, fitting like a sphere over all the rest.

That, however, would be a dead universe. In fact, it was a corollary of this whole view of the world that ordinary motion up or down or in a straight line could only take place if there was something wrong—something displaced from its proper sphere. It mattered very much, therefore, that the various elements were not all in order but were mixed and out of place—for instance, some of the land had been drawn out above the waters, raised out of its proper sphere at the bottom, to provide habitable ground. On this land natural objects existed and, since they were mixtures, they might, for example, contain water, which as soon as it was released would tend to seek its way down to the sea. On the other hand, they might contain the element of fire, which would come out of them when they burned and would flutter and push its way upwards, aspiring to reach its true home. But the elements are not always able to follow their nature in this pure fashion—occasionally the fire may strike downwards, as in lightning, or the water may rise in the form of vapour to prepare a store of rain. On one point, however, the law was fixed: while the elements are out of their proper spheres they are bound to be unstable—there cannot possibly be restfulness and peace. Woven, as we find them, on the surface of the globe, they make a mixed and chancy world, a world that is subject to constant mutation, liable to dissolution and decay.

It is only in the northern hemisphere that land emerges, protruding about the waters that cover the rest of the globe. This land has been pulled up, out of its proper sphere, says Dante—drawn not by the moon or the planets or the ninth sky, but by an influence from the fixed stars, in his opinion. The land stretches from the Pillars of Hercules in the west to the Ganges in the east, from the Equator in the south to the Arctic Circle in the north. And in the centre of this whole habitable world is Jerusalem, the Holy City. Dante had heard stories of travellers who had found a great deal more of the continent of Africa, found actual land much farther south than he had been taught to consider possible. As a true rationalist he seems to have rejected "fables" that contradicted the natural science of his time, remembering that travellers were apt to be liars. The disproportionate amount of water in the world and the unbalanced distribution of the land led to some discussion of the whereabouts of the earth's real centre. The great discoveries, however, culminating in the unmistakable discovery of America, provoked certain changes in ideas, as well as a debate concerning the possibility of the existence of inhabited countries at the antipodes. There was a growing view that earth and water, instead of coming in two separate circles, the one above the other, really dovetailed into one another to form a single sphere.

All this concerns the sublunary region; but there is

another realm of matter to be considered, and this, as we have already seen, comes under a different polity. The skies are not liable to change and decay, for they—with the sun, the stars and the planets—are formed of a fifth element, an incorruptible kind of matter, which is subject to a different set of what we should call physical laws. If earth tends to fall to the centre of the universe, and fire tends to rise to its proper sphere above the air itself, the incorruptible stuff that forms the heavens has no reason for discontent—it is fixed in its congenial place already. Only one motion is possible for it—namely, circular motion—it must turn while remaining in the same place. According to Dante there are ten skies, only the last of them, the Empyrean Heaven, the abode of God, being at rest. Each of the skies is a sphere that surrounds the globe of the earth, and though all these spheres are transparent they are sufficiently tangible and real to carry one or more of the heavenly bodies round on their backs as they rotate about the earth—the whole system forming a set of transparent spheres, one around the other, with the hard earth at the centre of all. The sphere nearest to the earth has the moon attached to it, the others carry the planets or the sun, until we reach the eighth, to which all the fixed stars are fastened. A ninth sphere has no planet or star attached to it, nothing to give visible signs of its existence; but it must be there, for it is the *primum mobile*—it turns not only itself but all the other spheres or skies as well, from east to west, so that once in twenty-four hours the whole celestial system wheels round the motionless earth. This ninth sphere moves more quickly than any of the others, for the spirits which move it have every reason to be ardent. They are next to the Empyrean Heaven.

In the system of Aristotle the spheres were supposed to be formed of a very subtle ethereal substance, moving more softly than liquids and without any friction; but with the passage of time the idea seems to have become coarsened and vulgarised. The successive heavens turned into glassy or crystalline globes, solid but still transparent, so that it became harder for men to keep in mind the fact that they were frictionless and free from weight, though the Aristotelian theory in regard to these points was still formally held.

The original beauty of this essentially Aristotelian system had been gravely compromised, however, by the improvements which had been made in astronomical observation since the time when it had been given its original shape; for even in the ancient world astronomy afforded a remarkable example of the progress which could be achieved in science by the sheer passage of time—the accumulating store of observations and the increasing precision in the recordings. Early in the Christian era, in the age of Ptolemy, the complications had become serious, and in the middle ages both the Arabs and the Christians produced additions to the intricacies of the system. The whole of the celestial machinery needed further elaboration to account for planets which, as viewed by the observer, now stopped in the sky, now turned back on their courses, now changed their distance from the earth, now altered their speed. However irregular the motion of the planets might seem to be, however curious the path that they traced, their behaviour must somehow or other be reduced to circular, even uniform circular motion—if necessary to a complicated series of circular motions each corrective of the other. Dante explains how Venus goes round with the sphere which forms the third of the skies, but as this does not quite correspond to the phenomena, another sphere which revolves independently is fixed to the sphere of the third sky, and the planet rides on the back of the smaller sphere (sitting like a jewel there, says Dante), reflecting the light of the sun. But writers varied on this point, and we meet the view that the planet was rather like a knot in a piece of wood, or represented a mere thickening of the material that formed the whole celestial sphere—a sort of swelling that caught the sunlight and shone with special brilliance as a result.

Writers differed also on the question whether the whole of the more elaborate machinery—the eccentrics or epicycles—as devised by Ptolemy and his successors, really existed in the actual architecture of the skies, though the theory of the crystalline spheres persisted until the seventeenth century. Since the whole complicated system demanded eighty spheres, some of which must apparently intersect one another as they turned round, some writers regarded the circles and epicycles as mere geometrical devices that formed a basis for calculation and prediction. And some men who believed that the nine skies were genuine crystalline spheres might regard the rest of the machinery as a mathematical way of representing those irregularities and anomalies which they knew they were unable properly to explain. In any case, it was realised long before the time of Copernicus that the Ptolemaic system, in spite of all its complications, did not exactly cover the phenomena as observed. In the sixteenth and seventeenth centuries we shall still find people who will admit that the Ptolemaic system is inadequate, and will say that a new one must be discovered, though for understandable reasons they reject the solution offered by Copernicus. Copernicus himself, when he explained why his mind had turned to a possible new celestial system, mentioned amongst other things the divergent views that he had found already in existence. The Ptolemaic system would be referred to as the Ptolemaic hypothesis, and we even find the Copernican theory described by one of its supporters as "the revision of the hypotheses." Many of us have gone too far perhaps in imagining a cast-iron Ptolemaic system, to the whole of which the predecessors of Copernicus were supposed to be blindly attached.

Finally, according to Dante, all the various spheres are moved by Intelligences or Spirits, which have their various grades corresponding to the degrees of nobility that exist in the physical world. Of these, the lowliest are the angels who move the sphere of the moon; for the moon is in the humblest of the heavens; she has dark spots that show her imperfections; she is associated with the servile and poor. (It is not the moon but the sun which affords the material for romantic poetry under this older system of ideas.) Through the various Intelligences operating by means of the celestial bodies, God has shaped the material world, only touching it, so to speak, through intermediaries. What He created was only inchoate matter, and this was later moulded into a world by celestial influences. Human souls, however, God creates with His own hands; and these, again, are of a special substance—they are incorruptible. Even now, long after the creation, the heavens still continue to influence the earth, however, says Dante—Venus affecting lovers, for example, by a power that comes not from the sphere but from the planet itself, a power that is actually transmitted by its rays. The Church had struggled long against the deterministic implications of astrology, and was to continue the conflict after the time of Dante, though means were already being adopted to reconcile astrology with the Christian teaching concerning free will. Dante said that the stars influence the lower dispositions of a man, but God has given all men a soul by which they can rise above such conditioning circumstances. Occasionally one meets even with the opposite view—that the stars can influence only for good, and that it is man's own evil disposition that is responsible if he turns to sin. Those who attacked astrology often took the line that the observation of the paths of the heavenly bodies was not sufficiently accurate as yet to allow of detailed predictions. Astrologers themselves, when their prophecies were found to be inaccurate, would blame the faultiness of astronomical observation rather than the defects of their own supposed science. The controversy between the supporters and the opponents of astrology could be turned into a channel, therefore, in which it became an argument in a circle. It would appear to be the case that astrology, like witch-burning, was considerably on the increase in the sixteenth and seventeenth centuries, in spite of what we say about the beginning of modern times.

In this whole picture of the universe there is more of Aristotle than of Christianity. It was the authority of Aristotle and his successors which was responsible even for those features of this teaching which might seem to us to carry something of an ecclesiastical flavor—the hierarchy of heavens, the revolving spheres, the Intelligences which moved the planets, the grading of the elements in the order of their nobility and the view that the celestial bodies were composed of an incorruptible fifth essence.

Indeed, we may say it was Aristotle rather than Ptolemy who had to be overthrown in the sixteenth century. It was necessary to make a great advance on the general scientific teaching of Aristotle before the world could be in a position to do justice to the Copernican hypothesis. Once again, this ancient teacher, perhaps by reason of the merit and the very power of his intellectual system, comes to appear as an obstruction to the progress of science.

The great work of Copernicus, *De Revolutionibus Orbium,* was published in 1543, though its author would appear to have been working upon it and elaborating his system since the early years of the century. It has often been pointed out that he himself was not a great observer, and his system was not the result of any passion for new observations. This passion came into astronomy later in the century, particularly with Tycho Brahe, who himself always refused to become a follower of Copernicus, and who amongst other things introduced the practice of observing planets throughout the whole of their courses, instead of just trying to pick them out when they happened to be at special points in their orbits. It was even true that Copernicus trusted too much to the observations that had been handed down by Ptolemy himself from the days of antiquity. In one of his writings he criticises a contemporary for being too sceptical concerning the accuracy of Ptolemy's observations. The later astronomer, Kepler, said that Copernicus failed to see the riches that were within his grasp, and was content to interpret Ptolemy rather than nature. It seems to have been one of his objects to find a new system which would reconcile all recorded observations, and a disciple of his has described how he would have all these before him as in a series of catalogues. It is admitted, however, that he fell into the mistake of accepting the bad observations and the good without discrimination. One modern writer has pointed out that since he was putting forward a system which claimed to account for the same phenomena as were covered by the Ptolemaic theory, he may have been wise in not laying himself open to the charge that he was doctoring Ptolemy's observations in order to fit them to his hypothesis. Through his trust in the ancient observations, however, he allowed himself to be troubled by irregularities in the sky which did not really exist; and in one or two ways he produced needless elaborations which were calculated to hinder the acceptance of his system.

If we ask, then, why he was moved to attempt a new interpretation of the skies, he tells us that he was disturbed by the differences of opinion that had already existed amongst mathematicians. There is evidence that one matter of actual observation gave him some bewilderment—he was puzzled by the variations he had observed in the brightness of the planet Mars. This was the planet which during the next century caused great difficulty to astronomers and led to most remarkable developments in astronomy. Copernicus's own system was so far from answering to the phenomena in the case of Mars that Galileo in his main work on this subject praises him for clinging to his new theory though it contradicted observation—contradicted in particular what could be observed in the behaviour of Mars. It would appear that Copernicus found a still stronger stimulus to his great work in the fact that he had an obsession and was ridden by a grievance. He was dissatisfied with the Ptolemaic system for a reason which we must regard as a remarkably conservative one—he held that in a curious way it caused offence by what one can almost call a species of cheating. Ptolemy had pretended to follow the principles of Aristotle by reducing the course of the planets to combinations of uniform circular motion; but in reality it was not always uniform motion about a centre, it was sometimes only uniform if regarded as angular motion about a point that was not the centre. Ptolemy, in fact, had introduced the policy of what was called the equant, which allowed of uniform angular motion around a point which was not the centre, and a certain resentment against this type of sleight-of-hand seems to have given Copernicus a special urge to change the system. That he was in earnest in his criticism of Ptolemy's device is shown both in the system he himself produced, and in the character of certain associated ideas that gave a strong bias to his mind.

A further point has been noted sometimes. It can best be explained, perhaps, if we imagine a competent player who stares at the draught-board until a whole chain of his rival's draughts seem to stand out in his mind, plainly asking to be removed by a grand comprehensive stroke of policy. An observer of the game can often be sensible of the way in which the particular pieces glare out at one —so many black draughts which are waiting to be taken as soon as the white opponent can secure a king. It would seem to have been the case that a mind so geometrical as that of Copernicus could look at the complicated diagram of the Ptolemaic skies and see a number of the circles which cried out to be removed provided you had a king to take them with—all of them would cancel out as soon as it occurred to you to think of the earth as being in motion. For, if the ancients ignored the fact that they—the spectators of the skies—were moving, it was inevitable that to each of the heavenly bodies they should have imputed an additional, unnecessary, complicating motion—sun, planets and stars had to have a tiresome extra circle in the diagram—and this in every case would be referable to the same formula, since it corresponded to what ought to have been each time the motion of the earth. As a geometer and a mathematician Copernicus seems to have been struck by the redundancy of so many of the wheels.

Finally in this connection it is necessary to remember the way in which Copernicus rises to lyricism and almost to worship when he writes about the regal nature and the central position of the sun. He would not stand alone if he proved to have been stimulated to genuine scientific enquiry by something like mysticism or neo-platonic sentiment. He held a view which has been associated with Platonic and Pythagorean speculations to the effect that immobility was a nobler thing than movement, and this affected his attitude to both the sun and the fixed stars. Many factors, therefore, combined to stimulate his mind and provoke him to a questioning of the ancient system of astronomy.

He had passed a number of years in Italy at one of the most brilliant periods of the Renaissance; and here he had learned something of those Platonic-Pythagorean speculations which had become fashionable, while acquiring, no doubt, some of the improved mathematics which had resulted from the further acquaintance with the achievements of antiquity. His reverence for the ancient world is illustrated in the way he always spoke of Ptolemy; and, having seen reason to be dissatisfied with the prevailing condition of things in astronomy, he tells us that he went back to study what previous writers had had to say on the whole question. Once again, as in the case of the theory of impetus, the new development in science was assisted by hints from ancient writers, and was stimulated by the differences of opinion that had already existed in antiquity. Some later medieval writers like Nicholas of Cusa had encountered the suggestion that the earth might be in motion, and had been willing to entertain the idea; but nobody had troubled to work out the details of such a scheme, and up to the time of Copernicus the heliocentric theory had never been elaborated mathematically in order to see whether it would cover and explain the observed phenomena in the competent way in which the Ptolemaic system had proved able to do. Only the Ptolemaic theory had hitherto possessed the advantage which the modern world would prize—the merit of having been established in a concrete way, with the demonstration that it fitted the facts (on the whole) when applied to the phenomena in detail. Copernicus may have been assisted somewhat by a view transmitted to the middle ages by Martianus Capella, which regarded just the two planets, Mercury and Venus, as going round the sun. These two planets, lying between the earth and the sun and always observed in close proximity to the latter, had long presented special problems to those people who tried to regard them as going round the earth.

Wherever he found the hint, Copernicus made it his real task to uncover the detailed workings of the skies under the new hypothesis and to elaborate the mathematics of the scheme. His own theory was only a modified form of the Ptolemaic system—assuming the same celestial machinery, but with one or two of the wheels interchanged through the transposition of the rôles of the earth and the sun. He made it a little harder for himself by trying to work all the observed movements of the planets into a more genuine system of uniform circular motion—uniform in respect of the centre of the circle, without any conjuring-tricks with equants. He had to use the old complicated system of spheres and epicycles, however, though he could claim that his hypothesis reduced the total number of wheels from eighty to thirty-four. Although some doubt has been expressed (and he himself declared the matter to be outside his concern) he appears to have believed in the actual existence of the rotating orbs—the successive crystalline heavens—and at any rate the astronomer Kepler considered this to have been the case. It was a disadvantage of his system that it was not quite heliocentric after all—the earth did not describe an exact circle with the sun as its centre, and, in fact, all the movements of the skies were reckoned not from the sun itself, but from the centre of the earth's orbit, which came somewhat to the side. This was significant, as it infringed the old doctrine that there must be a core of hard matter somewhere on which the other things actually hinged and turned—something more than a mere mathematical point to serve as the hub of the universe.

Since the older Ptolemaic theory had approximately accounted for the phenomena, while the Copernican system itself only accounted for them approximately, much of the argument in favour of the new hypothesis was to turn on the fact of its greater economy, its cleaner mathematics and its more symmetrical arrangement. Those who were unable to believe in the motion of the earth had to admit that for calculation and prediction the Copernican theory provided a simpler and shorter method. Whereas on the older view the fixed stars moved round in one direction at a speed which seemed incredible, while the planets largely turned in the opposite way and often seemed to be at sixes and sevens with the sun, it now appeared that the motion was all in the same direction—the earth and the planets, duly spaced and all in order, swung in the same sweeping way around the sun, the time of their orbits being related to their respective distances from the latter. Only thirty-four spheres or circles were needed instead of eighty, as we have seen. And by a simple daily rotation of the earth upon its axis you were saved from the necessity of making all the skies undertake a complete revolution every twenty-four hours.

On the other hand, some of what we might regard as the beautiful economy of the Copernican system only came later—for example, when some of Copernicus's own complications and encumbrances had been taken away. And if from a purely optical standpoint or from the geometer's point of view the new hypothesis was more economical, there was another sense in which it was more prodigal; because in respect of the physics of the sixteenth century it left a greater number of separate things that required separate explanation. In any case, at least some of the economy of the Copernican system is rather an optical illusion of more recent centuries. We nowadays may say that it requires smaller effort to move the earth round upon its axis than to swing the whole universe in a twenty-four hour revolution about the earth; but in the Aristotelian physics it required something colossal to shift the heavy and sluggish earth, while all the skies were made of a subtle substance that was supposed to have no weight, and they were comparatively easy to turn, since turning was concordant with their nature. Above all, if you grant Copernicus a certain advantage in respect of geometrical simplicity, the sacrifice that had to be made for the sake of this was nothing less than tremendous. You lost the whole cosmology associated with Aristotelianism—the whole intricately dovetailed system in which the nobility of the various elements and the hierarchical arrangement of these had been so beautifully interlocked. In fact, you had to throw overboard the very framework of existing science, and it was here that Copernicus clearly failed to discover a satisfactory alternative. He provided a neater geometry of the heavens, but it was one which made nonsense of the reasons and explanations that had previously been given to account for the movements in the sky. Here, as in the case of the doctrine of impetus, it was necessary to go farther still and complete the scientific revolution before one could squarely meet the criticisms to which the new hypothesis was liable. Kepler was right, therefore, when he said that Copernicus failed to see how rich he was, and erred by remaining too close to the older system of Ptolemy.

This point becomes clear when Copernicus tries to meet the objections to his hypothesis, and particularly when he has to show how the celestial machinery could possibly work, supposing his geometrical design to be correct. We are all aware that, when two trains are alongside, it is difficult to tell whether it is ours or the other one that is moving; and this purely optical relativity of motion must have long been realised, for otherwise it could never have occurred to either the ancient world or the middle ages that it was possible to discuss whether it was the earth or the sun which moved. Anybody may grant the point concerning the relativity of motion for the sake of argument, but still this does not decide for us the crucial question—it does not tell us which of the trains it is that is actually moving. In order to discuss this question at all Copernicus had to go into the further problem of the nature and cause of movement—had to move from geometry and from the problem of the mere pattern of the skies to issues which belonged rather to physics. If one asked Copernicus why the earth and the heavenly bodies moved, he would answer: Because they were spherical or because they were attached to spherical orbs. Put a sphere anywhere in space and it would naturally revolve—it would turn without needing anybody to turn it—because it was the very nature of the sphere to rotate in this way. Whereas Aristotle had made movement depend on the total nature of the celestial bodies as such, it has been pointed out that Copernicus observed with something of the geometer's eye; for in his argument the nature of the body was decided purely by the geometrical shape, and the movement merely depended upon sphericity. All bodies, furthermore, aspired to become spheres—like water forming drops—for the simple reason that the sphere represented the perfect shape. Gravity itself could belong to the sun and moon as well as to the earth—could belong to anything spherical—since it represented the tendency of all parts of a body to come together and consolidate themselves in the form of a sphere.

In fact, from a certain point of view the actual movement of the earth falls into place as almost an incidental matter in the system of Copernicus, which, viewed geometrically, as I have already said, is just the old Ptolemaic pattern of the skies with one or two of the wheels interchanged and one or two of them taken out. If one stares long at the new system, it seems to be a set of other characteristics that begin to come out into relief, and these have the effect of putting Copernicus into remarkable contrast with both the older world and modern times. Not only is Copernicus prodded and pressed into overturning the old system by a veritable obsession for uniform circular motion (the point on which he thought that Ptolemy had shuffled and faked), but in facing the two biggest problems of his system, the dynamics of it and the question of gravitation, he gives an unexpected turn to the discussion by a very similar obsession in regard to the sphere as the perfect shape. It is amazing that at this early date he had even faced these colossal issues—matters concerning which his successors went on fumbling down to the time of Galileo and even Newton. But on the first big question: What were the dynamics of the new system?—By what physical law did bodies move in the Copernican way?—his answer was neither the doctrine of impetus as such nor the modern law of inertia, but the view that spherical bodies must turn —the earth itself according to this principle could not help turning. If you take the other great problem: Now that the earth is no longer the centre of the universe, what about the whole Aristotelian theory of gravity?—Copernicus jumps to what in one aspect is the modern view: that not only the earth but other bodies like the sun and the moon have gravity. But he ties the whole notion to the same fundamental principle—the tendency of all things to form and consolidate themselves into spheres, because the spherical shape is the perfect one. That is why his synthesis is so colossal—he not only replaces Ptolemy's astronomy but he attacks Aristotle's physics on matters of the profoundest principle. And the passion which is the motor behind everything is connected with what might seem to us almost an obsession for circularity and sphericity—one which puts the ancient Ptolemy into the shade. When you go down, so to speak, for the third time, long after you have forgotten everything else in this lecture, there will still float before your eyes that hazy vision, that fantasia of circles and spheres, which is the trade-mark of Copernicus, the very essence of undiluted Copernican thought. And, though it had influence in the sixteenth century we must note that it never came in that significant way into the ideas of the seventeenth century or into the science of the modern world.

In general, it is important not to overlook the fact that the teaching of Copernicus is entangled (in a way that was customary with the older type of science) with concepts of value, teleological explanations and forms of what we should call animism. He closes an old epoch much more clearly than he opens any new one. He is himself one of those individual makers of world-systems, like Aristotle and Ptolemy, who astonish us by the power which they showed in producing a synthesis so mythical—and so irrelevant to the present day—that we should regard their work almost as a matter for æsthetic judgment alone. Once we have discovered the real character of Copernican thinking, we can hardly help recognising the fact that the genuine scientific revolution was still to come.

Within the framework of the older system of ideas, Copernicus was unable to clinch his argument. To the old objection that if the earth rotated its parts would fly away and it would whirl itself into pieces, he gave an unsatisfactory answer—he said that since rotation was for the earth a natural movement, the evil effects could not follow, for the natural movement of a body could never be one that had the effect of destroying the nature of that body. It was the argument of a man who still had one foot caught up in Aristotelianism himself, though perhaps precisely because it seems to be archaic to us it was more appropriate to those conservative people whose objections required to be met in the sixteenth century. When it was said that if the world was rushing from west to east (in the way that Copernicus suggested) the air would be left behind and there would be a continual wind from east to west, he still answered somewhat in the terms of the ancient physics, and said the air must go round with the globe because of the earthiness which the atmosphere itself contained and which rotated in sympathy with everything else that was earthy. He was not much more successful when he tried to take the argument about the earth whirling itself to pieces, and turn it against the possible critics of his theory, saying that the skies themselves, if they were to rotate so quickly as was assumed, would be broken into fragments by the operation of the very same laws. The skies and the heavenly bodies, as we have seen, were supposed to have no weight—on the Aristotelian theory they were not regarded as being subject to the operation of what we now call centrifugal force. Galileo himself apparently made mistakes when trying to meet the argument that the world would fly to pieces if it were turning on its own axis. That whole question of centrifugal force proved to be a serious obstruction to the acceptance of the Copernican system in the sixteenth century, and was only made manageable by the work of men like Huygens, whose writings appeared more than a hundred years after *De Revolutionibus Orbium*. The truth was that Ptolemy in ancient times had rejected the hypothesis of the movement of the earth, not because he had failed to consider it, but because it was impossible to make such an hypothesis square with Aristotelian physics. It was not until Aristotelian physics had been overthrown in other regions altogether that the hypothesis could make any serious headway, therefore, even in modern times.

If Copernicus is so far from being a representative of the modern outlook, his case may serve to remind us how often in the mentality of the Renaissance we meet with features which we today would have to regard as archaic. It has been pointed out that men in those days who did not mean to be mystics, lost in the supernatural, were perhaps rather that we might call surrealists, projecting the fabulous animals of antiquity and the products of their own imagination on to what they thought was the concrete world. Now, therefore, more than in the middle ages, "the basilisk, the Egyptian phœnix, the gryphon, the salamander, again came in to their own." There was some imposing philosophical thought in this period, and if much of this was concentrated on the problem of the soul or the question of the dignity of man, much of it represented also an attempt to see the whole of nature as a single self-explanatory system. The object was to eliminate transcendental influences—the activity of spirits and demons working on the world from outside, or the capricious intervention of God Himself—and to seek the explanations of all phenomena within the actual system of nature, this system being regarded as self-sufficient and as working under the government of law. There was even a growing insistence that the phenomena of nature should be more carefully observed and that data should not simply be accepted on the authority of ancient writers. This whole movement has its place in the history of science, therefore, and made its contribution to the story that we have to examine, though later it was to become an obstruction and in some respects it now seems less rational to us than the scholasticism of the middle ages.

Under the cover of the revival of antiquity, ancient forms of occultism, Jewish cabalistic speculation, Arabian magical arts, and alchemical mystification came to be mixed with the ingredients of philosophy. These influences helped to bring back views of the universe that were older than Aristotle—forms of what might be called pan-psychism and astro-biology and animism. If the belief in astrology and witchcraft and alchemical speculation went on increasing from this time, they were encouraged by a fashionable philosophy and a prevailing intellectual tendency—so that what we could call magic belonged not merely to popular superstition but to the high-browism of the age. The attempt to achieve a unified idea of the cosmos was in fact perhaps bound to result in something fanciful, in view of the imperfection of the data then available.

The Renaissance naturalists might be anxious to reject miracles, but like Pomponazzi they would still believe that certain plants or stones might bring rain, that animals could prophesy, or that a statue might perspire to announce a grand event. These things could be accepted because they were regarded as amongst the verified data in the universe, and it could even be held that prayers might conjure away a storm not by any intervention of God but by the displacements they produced in the atmosphere. It was possible, therefore, to apply one's mind solely to what one regarded as nature, but to see nature itself as magical. And some who were determined to apply criticism were unable to reject the evidence for astrology—some even set out to purify it from superstition, without conceiving that it was itself in any way a fraud.

On this system, it was important to attach phenomena to their causes, and a real chain of cause and effect was taken for granted, though there might be no rigorous distinction between material and mental phenomena, between mechanical and occult activity. To a certain degree, causes were sought by hunting out analogies and mystical correspondences between things—imagining the stars as male or female, or as hot or cold, and giving them special affinities with minerals or with parts of the human body, so that the whole universe seemed sometimes to be a universe of symbols. The action of the magnet seems to have influenced these thinkers very much. It appears even to have been taken as a typical example of the way things work in nature, so that there was a search for secret magical sympathies between objects. The fact that plants had sex, and the manner in which light was diffused throughout the world of nature, had a typical or symbolical significance in somewhat the same way. At the same time, the thinkers took into their survey various phenomena connected with telepathy, hypnotism, etc., which have become interesting to us again in the twentieth century. They would show, for example, how, by bringing methods of suggestion into play, one might discover that a man was a thief.

All such kinds of thinking culminated, perhaps, in

Chapter 3

The Study of the Heart down to William Harvey

THE RENAISSANCE in one of its important aspects sees the culmination of the long medieval process of first recovering, then translating and finally assimilating, the scholarly and scientific writings of antiquity. This in itself may have added no new ingredients to a civilisation that was Græco-Roman in character; but the exhilaration which it produced was very great, and, together with the blossoming of urban life in city-states, it seemed to bring liveliness to the intellect. The world now became aware, moreover, that Aristotle had not been without his rivals in the ancient world; and the confrontation of rival explanations, rival systems, provided significant issues that men were compelled to decide for themselves. The discovery of the new world, and the beginning of a close acquaintance with tropical countries, released a flood of new data and a mass of descriptive literature which itself was to have stimulating effects. The essential structure of the sciences was not changed—the scientific revolution was still far off—but the Renaissance has perhaps a greater significance in the biological field than it appears to have had in the case of physics. Devices that were associated with printing—such things as woodcuts and copperplate engravings—put new instruments at the disposal of the scientific teacher. At last one could be sure that drawings and diagrams would be copied and multiplied with accuracy; and this, together with printing itself, made accurate recording and scientific intercommunication more practicable. Vesalius, who is of some importance in our story, is particularly associated with this development—significant in his use of illustrations and in the naturalistic character that he gave them, so different from the conventionality of the medieval drawings.

There is a sense in which the art of fifteenth-century Italy may claim perhaps to stand as a chapter in the history of the rise of modern science. The practitioners and theorists who insisted that painting was a branch of knowledge were not merely bidding for a higher status in the world; and from Alberti to Leonardo da Vinci they insisted, for example, on the importance of mathematics, or even claimed it as the primary qualification of the artist. Apart from the study of optics and perspective, of geometry and proportion, great significance was attached to anatomy; and here the artist could observe for observation's sake—observe without all the preoccupations of the medical student, anxious to appropriate the whole Galenic theory of the human body. It has been considered the virtue of Masaccio, after the end of the first quarter of the century, that he had painted his figures in the round, and not merely as flat things on the canvas, and for this reason it was said of him that he was the first man to reproduce things as they really are. From this time, the Florentine school of painters distinguished itself by the intentness with which it concentrated on the purpose of naturalistic representation—especially where the human figure was concerned—a purpose which, from a certain point of view, might be regarded as quasi-scientific. The goldsmith's workshop and the artist's studio would seem to have been forerunners of the modern laboratory; the very materials of the painter were the subject of research and experiment; in other ways there must have been close relations with the artisan—all very different from the natural philosopher producing scientific theories at his desk. Indeed, the artist, the artisan and the natural philosopher seem to be compounded together in the evolution of that modern figure, the natural scientist. In the fifteenth century the artist was often a technician—often the inventor of gadgets—an expert in machines, hydraulics and fortifications. Both before Leonardo da Vinci and after him, one painter after another is commissioned as a military engineer. Amongst the Florentine painters, and particularly the more mediocre ones, it would appear that at times the science prevailed over the art itself—the pictures showed off a knowledge of muscles or virtuosity in the handling of problems of perspective. It has been pointed out that the men who were drawn to the artist's studio in the fifteenth century are just those who—through the same temperament—were to be attracted to the workshop of Galileo in the seventeenth century. Above all, the art of empirical observation must have been greatly developed. And the artists, much as they owed to the ancient world, were the first to cry out against mere subservience to authority—the first to say that one must observe nature for oneself.

Concerning some of the writers of the sixteenth century, it has been discovered that, though they talked of the importance of seeing things with one's own eyes, they still could not observe a tree or a scene in nature without noticing just those things which the classical writers had taught them to look for. When Machiavelli pretended to be drawing his conclusions from contemporary political events, he would still produce maxims drawn from one ancient thinker or another—he may have thought that he was making inferences from the data in front of him, but in reality he was selecting the data which illustrated the maxims previously existing in his mind. Similarly, the historical student, confronted with a mass of documentary material, has a kind of magnet in his mind which—unless he is very careful—will draw out of that material just the things which confirm the shape of the story as he assumed it to be before his researches began. In the later middle ages men realised that in the last resort everything depended on observation and experience, on dissection and experiment—so much so that in the fifteenth century we can find a man who claims to be communicating the result of his own experience and experiments, when in reality we now know that he was transcribing passages from the work of another writer. But though dissection was being practised to an increasing degree it produced little result—men only observed the things which the ancient writer Galen had taught them to look for.

Apart from all this, it seems to be true even today that when conclusions have been reached in science or in history, it is normal to incorporate these into a realm of "established facts"; after which they come to be transcribed from one book to another, as though now you had reached the end of the matter and on these topics the mind could rest. Students of history do not themselves authenticate every square inch of the history that they handle, and an outsider may wonder whether even scientific students authentically re-discover by independent experiment every particle of that science which they have built up in their minds. Now, much of the dissection that was being done at the opening of modern times was not a form of what we should call research at all, but was rather a demonstration before a class—not a process of discovery, but a way of teaching and illustrating established facts. Its very purpose was to communicate to students the truths that were in Galen, and a mere assistant would apparently conduct the actual work of dissection while the lecturer read the appropriate passage from the book. These poor creatures knew that Galen was a much greater artist in the work of dissection than they would ever be, and they took no end of pride in themselves if the result came out as Galen said it ought to do—a matter not at all easy, especially as Galen used apes, for example, instead of human bodies when he conducted his dissections. When in the field of anatomy an original mind emerged in the second quarter of the sixteenth century in the person of Vesalius, even he, finding that he differed from Galen, said (like others) that at first he could not believe his own eyes.

The Renaissance, then, brought more translations of the scientific work of the Greeks—in botany, for example—translations even of Galen himself. It brought conflicts concerning the authorities of the ancient world—disputes in the universities between Aristotle and Galen, for example, some of them turning on the problem of the function and the activity of the heart. There was also conflict between the so-called Arabians and the so-called Greeks—the former taking their Galen through Arabian transmission while the latter went direct to the original Greek—but it would appear that this controversy did not turn on the question of the heart. Finally, the Renaissance brought a greater insistence on observation and a refinement of observational skill. It is perhaps not an accident that the first branch of science transformed by improved observation was that of anatomy, the science of the painter, the one restored by Vesalius, in whom the mind of the artist and the mind of the scientist seem almost to have been fused into one.

Even in regard to the question of the heart, the influence of antiquity was not without its place in the story of the scientific revolution. William Harvey, whose demonstration of the circulation of the blood is the crowning-point of this episode, was Aristotelian in many respects, and was associated with the Aristotelian university of Padua. Like Aristotle he regarded the heart as the central organ of the body—the principal feature—so that on the subject of the heart he waxes so lyrical sometimes that he half-reminds us of Copernicus on the subject of the sun. Galen rather emphasised the importance of the liver and regarded the veins as centering in the liver. The Aristotelian view fell into a corresponding extravagance and said that even the nerves ran from the heart, which had a special importance as being the seat of the emotions.

The ancients had practised dissection, and Galen, besides dissecting animals, had studied human skeletons, and made experiments on living creatures. In fact, it was from Galen that the medical students—in a university like that of Padua, for example—had learned to be in advance of the other scientists in their general attitude to experiment. There had always been two pitfalls in the practice of ordinary dissection, however. First of all, many conclusions had been reached, not from human but from animal dissection, and even Vesalius in the work which inaugurated modern anatomy had to resort to animals instead of human beings for certain parts of the body. Secondly—and this was a fault that still had to be pointed out by William Harvey in the seventeenth century—people made wrong inferences when they examined animals which had been bled to death, so that the arteries and the left ventricle of the heart were seen only after the blood had been drained out of them. Galen, however, had exposed the error of the ancient view that the arteries and the left side of the heart contained only air. At the same time he regarded arterial blood as being mixed with a sort of spirituous substance called *pneuma,* a life-principle analogous in some ways to air and in other ways to fire.

Galen, however, was responsible for an important heresy which flourished in the sixteenth century and which it was the particular virtue of William Harvey to dispose of in a final manner. He held that one kind of blood ran from the liver through the veins to all parts of the body to perform an ordinary nutrifying function, while a different kind of blood, mixed with vital spirits in the way that I have already mentioned, flowed out through the arteries to perform a more vivifying kind of work. There were a great many hurdles to leap over before the seventeenth century came to the modern view that the blood goes out in the arteries and then comes back to the heart through the veins, Galen himself having had no conception of anything more than a sort of ebb and flow that occurred in the veins and the arteries independently. In order to understand the difficulties it is necessary for us to distinguish between the various processes which the ancient teaching on this subject involved—the key-points in the system which had to be overthrown. Implied in the Galenical view, first of all, was a direct passage of air from the lungs to the heart, where it was supposed to prevent any excess of that heat which it was the function of the heart to supply. Secondly, on the same view the chief action of the heart took place on the diastole, the dilatation—the main process was the drawing of the blood into the heart, and not its expulsion. Finally, on the same view the venous blood was drawn initially into the right side of the heart, but some of it seeped through a thick dividing-wall called the septum into the left ventricle of the heart, and here it was purified and mixed with vital spirits, the new mixture passing finally into the arteries on its own motion. The pivot of the whole system—the mechanically essential part of the scheme—was the passage of the blood across the dividing-wall, the septum, into the left ventricle, whence it moved into the arteries.

Here we have a complex system of errors concerning

which it has to be noted that the doctrine was not only wrong in itself, but, until it was put right, it stood as a permanent barrier against physiological advance—for, indeed, nothing else could be right. It is another of those cases in which we can say that once this matter was rectified the way lay open to a tremendous flood of further change elsewhere.

Now an Arabian physician of the thirteenth century denied that there were either visible passages or invisible pores in the septum of the heart to admit of the transit of blood from one side to the other. He maintained that only via the lungs could the blood make its way from the right ventricle of the heart to the left. A Latin translation of his work was printed in 1547 but it did not contain his teaching on this subject, so that the developments which took place in Italy after this date must be regarded as part of an independent movement. In the later middle ages there had been a dissecting student in western Europe who had said that the passages across the septum were very hard to find. Even Leonardo da Vinci fell into the error of thinking that the blood passed through the septum, though it is possible that he had some doubt about it in later life. After him, there is at least one writer—Vesalius himself—who says that it shows the mighty power of God that He should be able to make blood pass through the apparent solidity of the septum. In fact, it is this man Vesalius who first casts doubt on the whole Galenical teaching in regard to the septum, though it must be noted that, while correcting Galen, after much hesitation, on this particular point, he still does not see the necessity of a totally new conception of the movement of the blood.

The year 1543, which saw the publication of Copernicus's great work and of the important translation of Archimedes, is a date of considerable significance in the scientific revolution, because it saw also the publication of the *magnum opus* of Vesalius, namely the *De Fabrica,* the work which stands as the foundation of modern anatomy. Vesalius had started as a fervent follower of Galen, but it appears that from 1538 he had begun to have more and more doubts; for he was the genuine kind of pioneer, who carried out his dissections himself, invented new instruments for the purpose or borrowed them from workers in other fields, and established new forms of technique—for example, in regard to the mounting of skeletons. We must not imagine, however, that because he had doubts about the passage of the blood across the septum of the heart this heresy disappeared from the world of scholarship; especially as he expressed his doubts very cautiously and was particularly diffident in his first edition of 1543, where, as he later confessed, he deliberately accommodated his results to conform to the teaching of Galen. His achievement, his methods and (as we have seen) his illustrations, make him an important turning-point in our story, in spite of what he retained of the old mentality; and it was unfortunate that he quitted the life of research while still in his thirties and became physician first in the army and then at the court of the Emperor Charles V. In spite of his acknowledged greatness, he did not realise that a new account of the movements of the heart and blood was now necessary—in general, he retained the Galenical teaching on these points.

That was the first step towards the work of William Harvey, and at the next stage of the argument a different issue was in question, namely, the problem of the part played by the lungs in the system and the movement of the blood. Leonardo da Vinci had attacked the old view that air passed from the lungs into the heart—in fact, he said that he had tried with a pump and that it was impossible to force air into the heart by that route. It was the successor of Vesalius in Padua, a man called Colombo, who, in a publication of 1559, correctly described what is called the smaller circulation, the passage of the blood from the right side of the heart into the lungs and thence into the left ventricle of the heart. Apart from this single point Colombo remained true to the older Galenic doctrine on this subject—namely, that the blood flowed to the outer parts of the body not only through the arteries but also through the veins—in other words, Colombo had no idea of the larger, general circulation of the blood. Earlier than this, the famous heretic, Miguel Serveto, had published his *Christianismi Restitutio* in 1553, a work which was almost totally destroyed, since it was equally offensive to Catholic, Lutheran and Calvinist; and here he had inserted an account of the passage of the blood from the heart to the lungs and then back again to the left ventricle of the heart. As doctrines were sometimes communicated and treatises written some years before their actual appearance in print, scholars have not quite agreed on the question whether Serveto's discovery really precedes the achievement in Padua. Nor is it known whether the one party owed any debt to the other.

Another Italian writer, Cesalpino, is of interest and is sometimes claimed by the Italians as the real discoverer of the general circulation of the blood, and if the claim had been genuinely established he would have stood well in advance of William Harvey, for it is argued that he established his conclusions by 1593 and that these were disclosed in a work posthumously published in 1606. He was a great disciple of Aristotle, and put forward many stimulating things in his defence of the Aristotelian system as against the system of Galen. But though he talked of something like a general circulation and even saw some passage of blood from the arteries into the veins at the extremities of each, it is not clear that he realised the normal flow of all the blood from the one to the other; and though he saw that the blood in the veins moved towards the heart, he did not seize upon the importance of the discovery. Certainly he did not demonstrate the circulation as Harvey was to do, or grasp the whole in a mighty synthesis. It is unfortunate that Cesalpino's name has been discussed so largely in reference to this particular controversy only; for the concentration upon this single side of his work has prevented people from recognising many other things of interest in his writings.

By this time, another great step towards the discovery of the circulation of the blood had been made by Fabricius, who in 1574 published a work describing certain valves in the veins. Though he may not have been the first to make the discovery, the identification of these was most significant because it was perceived that the valves operated to check only the outward passage of blood from the heart and through the veins—the passage to the hands and feet by that route, for example. Such being the case, one might have expected that Fabricius would have realised that the normal passage was in the opposite direction—inwards, towards the heart—and that the blood in the veins was to be regarded therefore as being on its return journey. The mind of Fabricius was so shaped to the teaching of Galen, however, that he missed the whole point of his own discovery, and produced an explanation which left the large question exactly where it had stood before. He said that the valves merely served the purpose of checking and delaying the flow of blood, lest it should run too copiously to the hands and feet and collect there to excess, carried down, so to speak, by its own weight. Fabricius, in fact, was conservative in regard to many things and was still convinced that air passed directly from the lungs into the heart. When it was really demonstrated that blood was carried into the lungs by that very same passage-way from the heart, along which men believed that air moved in the opposite direction from the lungs into the heart, it was still necessary for William Harvey to point out that two such movements in opposite directions could hardly be taking place along one and the same passage at the same time. Until the seventeenth century, therefore, a curious mental rigidity prevented even the leading students of science from realising essential truths concerning the circulation of the blood, though we might say with considerable justice that they already held some the most significant evidence in their hands.

It is only by glancing at the obstructions and hurdles which existed in the sixteenth century, and by watching these earlier stages of fumbling piecemeal progress, that we can gain some impression of the greatness of William Harvey, who early in the seventeenth century transformed the whole state of the question for ever by a few masterly strategic strokes. Francis Bacon noted that some scientific discoveries appear deceptively easy after they have been made. He called to mind certain propositions in Euclid which appeared incredible when you first heard them propounded, yet seemed so simple after they had been proved that you felt they were something that you had really known all the time. We who look at the story from the wrong side of the great transition—with the history inverted because we know the answer beforehand—are tempted to see Harvey's predecessors as foolish and *ipso facto* to miss the greatness of Harvey's own achievement. Yet once again we must wonder both in the past and in the present that the human mind, which goes on collecting facts, is so inelastic, so slow to change its framework of reference. The predecessors of Harvey had observed by cuts and ligatures the flow of the blood in the veins towards the heart—not towards the outer parts of the body as their theory always took for granted. But they were so dominated by Galen that they said the blood behaved irregularly when it was tortured by such experiments—rushing off in the wrong direction like a fluttering frightened hen.

Before dealing with the actual achievement of Harvey we may note that for some years he was at the university of Padua, where the chief of his predecessors, Vesalius, Colombo and Fabricius, had worked. It is not possible to close one's eyes to the fact that this whole chapter in the history of the study of the heart is primarily the glory of that university. The credit which is due to Italy in respect of this is greater than any which her patriotic historians could win for her, supposing they made good the claims they have made on behalf of Cesalpino. From the beginning to the end our story is connected with the university of Padua, and the attention of the historian should be directed there. Furthermore, both Copernicus and Galileo were at the same university at important periods of their lives; and apart from the splendour of these great names there were developments in that university which would justify the view that in so far as any single place could claim the honour of being the seat of the scientific revolution, the distinction must belong to Padua. The great current which had taken its rise in the scholasticism of fourteenth-century Paris passed particularly to the universities of north Italy, and in the sixteenth century it is there that the doctrine of the impetus is developed, at a time when the tradition and interests of Paris itself were moving away from that branch of study. To the humanists of the Renaissance, Padua was an object of particular derision because it was the hotbed of Aristotelianism; and it is one of the paradoxes of the scientific revolution that so important a part was played in it by a university in which Aristotle was so much the tradition and for centuries had been so greatly adored. Padua had certain advantages, however—it was a university in which Aristotle was studied largely as a preliminary to a course in medicine; for here medicine was the queen of the sciences, rather than theology, as was the case in Paris. As I have already mentioned, Galen had handed down to the medical students a regard, not merely for observation, but for actual experiment; and, further than this, his own writings had given an impulse in Padua to the conscious discussion of the experimental method. In any case, the Italian cities at this date had become highly secularised, and in Padua this had been particularly noticeable for a long time—noticeable even in political thought, as is evidenced by the work of Marsiglio. The interpretation of Aristotle in the university was largely concerned with his writings on the physical universe, and the study had long been carried out in collaboration with the medical faculty itself. Whereas the scholastic philosophers of the middle ages had assimilated Aristotle into their Christian synthesis, the Paduans made him a much more secular study—looking rather to the original Aristotle, naked—that is to say, without his Christian dress. Or, rather, perhaps one ought to remember that the Paduans were inclined to adopt this attitude because they were an Averroist university—seeing Aristotle in the light of the Arabian commentator, Averroes. Padua fell under the rule of Venice from 1404, and Venice was the most successfully anti-clerical state in Europe both at this time and for long afterwards. The freedom of thought enjoyed by Padua attracted the ablest men, not only from the whole of the Italian peninsula, but also from the rest of Europe—William Harvey himself being a conspicuous example of this. In the first volume of the *Journal of the History of Ideas* an article by J. H. Randall points out that we have exaggerated the importance of the new thought of the Renaissance—particularly the cult of Platonic-Pythagorean ideas—in the scientific revolution. There is a deeper continuity of history between the fifteenth and the seventeenth centuries in the conscious discussion of the scientific method in the university of Padua; and here again we have to note the way in which Aristotle was overthrown by the mere continuance and development of the process of commentating on Aristotle. It has been pointed out that fifteenth-century discussions of scientific method in Padua attacked the question of the purely quantitative (as against the older qualitative) mode of treatment; that in the sixteenth century the Paduans were doubting the old view that natural motion, in the case of falling bodies, for example, was the result of a tendency inherent in those bodies—they were asking whether it was not perhaps the effect of the exertion of a force. Towards the end of the sixteenth century they were questioning whether final causes ought to have any place in natural philosophy. They had extraordinarily clear views concerning scientific method, and Galileo, who arrived in that university just after some of the greatest controversies had taken place on that subject, inherited some of these points of method and used the same terminology when he discussed their character. It was in anatomy, however, that they achieved their most signal results as a school and as a tradition; and it is of considerable importance that William Harvey was initiated into that tradition; for one of the remarkable features of his work was not merely dissection and observation, but actual experiment. As late as 1670 we find an English work which describes Padua as "the Imperial University for Physic of all others in the world."

William Harvey continued and developed, not merely the dissections and observations, but also the kind of experiments which are to be seen in sixteenth-century Padua. He declared that he both learned and taught anatomy, "not from books but from dissection," and he combined the results of this with clinical observation and ingenious experiment. What is very remarkable in him is the comprehensive and systematic character of his investigation as a whole—not merely in ranging over so many of the operations and so much of the topography of the blood-circuit, but also in extending the comparative method so systematically over so large a range of creatures. After we have heard so much about the mistakes that had been made through dissecting apes and other animals instead of human beings, it is curious to find Harvey making it a matter of complaint that dissection was now too often confined to the human body—that insufficient attention was being given to the comparative method. An extraordinarily modern flavour attaches to his work as a result of the clearly mechanical nature of much of his enquiry and his argument; the importance that he gave to purely quantitative considerations; and the final cogency that he attributed to a piece of arithmetic. It is interesting to see him talking about the heart as "a piece of machinery in which though one wheel gives motion to another, yet all the wheels seem to move simultaneously." When he examined any anatomical feature he did not pretend immediately to deduce its function from an impression of its form and structure, but, once a hypothesis suggested itself, sought for the experiment which would mechanically confirm the idea he had conceived. Finally, though his book itself seems to lack order, it gives a clear account of the methods employed at the various points of the argument, and is remarkable as a comprehensive record of experiments made.

It seems to have been the valves which set Harvey thinking, possibly the valves at the entries or outlets of the heart itself, though very soon he was concerned with those valves in the veins which his teacher Fabricius had described and which he himself appears to have regarded as the stimulus to his enquiry. His book, *De motu cordis,* was published in 1628, but he himself wrote that "for more than nine years" he had been confirming his views "by multiple demonstrations." He still had to fight the old heresies, and we should make a mistake if we imagined that the discoveries made by those Paduan precursors of his who have been mentioned had as yet become common property. He attacks the view that the arteries take in air, and finds it necessary to point out that Galen himself had shown them to contain nothing but blood. He notes that when they are cut or wounded they neither take in air nor expel it in the way that the windpipe does when it has been severed. He still sets out to undermine the idea that blood crosses the septum of the heart—that septum, he says, is "of denser and more compact structure than any part of the body itself." If blood seeped through it, why did the septum need to have its own private supply through the coronary veins and arteries like the rest of the fabric of the heart? In any case, how could the left ventricle draw blood from the right when both contracted and dilated simultaneously? Similarly, he attacked the vexed question of the lungs—asked why something that had all the structure of a major blood-vessel should be supposed to have the function of carrying air from the lungs to the heart, while blood, on the other hand, was alleged to strain itself so laboriously through the solid septum of the heart. He made inferences from the structure of the vessels, made experiments to test the direction of the blood in them, and argued from their size that they must not merely carry blood necessary for the private nourishment of the lungs, but transmit all the blood through the lungs for the purpose of refreshing it, though he was unaware of the process of oxygenation which actually takes place. In addition, he used the comparative method, showing that creatures without lungs had no right ventricle of the heart —which confirmed his view that the right ventricle was connected with the transit to the lungs. He was able to show that in the embryo the blood took a shorter route from the right to the left ventricle of the heart—a route which ceased to be operative when the lungs came into action. Then he examined the fibrous structure of the heart, and showed that—contrary to the hitherto accepted view—its real activity consisted in its contraction and constriction—that is to say, in its systole, when it pumped out blood, and not in its diastole, when it had been alleged to suck it in. His actual description of the structure and the action of the heart may be said to rank as an admirable piece of artistry.

He showed that existing opinions on these subjects were neither plausible nor self-consistent; but though he was remarkable in his use of the comparative method he was not at his most original in the experimental field, where there appear even to have been devices, already put to use, which he did not employ. The revolution that he brought about was like the one which we have seen in the realm of mechanics, or the one that Lavoisier was to achieve in chemistry—it was due to the power of seeing the whole subject in a new framework and re-stating the issue in a way which made the problem manageable. It was due in fact to a kind of strategic sense which enabled his mind to seize on the point of crucial significance. Harvey's crowning argument is a simple piece of arithmetic, based on his estimate—a rough and inaccurate estimate—of the amount of blood which the heart sends through the body. It did not matter that his measurement was only a rough one—he knew that his conclusion must be right, even granting the largest margin of error that anybody might impute to him. The answer was clear to any mechanically-minded person who could really bring his attention to the point; and it rendered the rest of Harvey's evidence and argument merely subsidiary.

In regard to the capacity of the heart, he presents us with a sentence that can be left to speak for itself:

> What remains to be said upon the quantity and source of the blood which thus passes is of a character so novel and unheard of that I not only fear injury to myself from the envy of a few but I tremble lest I have mankind at large for my enemies, so much doth wont and custom become a second nature.

He found that in the space of an hour the heart would throw out more blood than the weight of a man, far more blood than could be created in that time out of any nourishment received. It was impossible to say where all that blood came from and where it could possibly go, unless one adopted the hypothesis that it went streaming through the whole body time after time in a continual circulation. Harvey followed that circulation from the left ventricle of the heart and round the body, showing how it explained the position of the valves of the heart, and accounted for the harder structure of the arteries, especially near the heart, where they had to bear the shock of each propulsion. He was now able to demonstrate why so often in bodies the blood had drained away from the arteries but not the veins, and he could give a more satisfactory reason for those valves in the veins which left the way to the heart open but prevented the return of the blood to the outer branches of the veins. The one link in the chain, the one part of the circulation, which he failed to trace was the passage of the blood from the outermost ramifications of the arteries to the outlying branches of the veins. The connection here could only be discovered with a microscope, and was made good in 1661 when Malpighi announced his identification of what are called the capillaries in the almost transparent lungs of a frog.

It seems to have taken between thirty and fifty years for Harvey's work to secure acceptance, though his arguments would perhaps seem more cogent to us today than those of any other treatise that had been written up to this period; since, though he held some of the unsatisfactory speculative views which were current at that time—such as a belief in vital spirits—his arguments never depended on these, and his thesis was so mechanically satisfying in itself that it helped to render them meaningless and unnecessary in future. Descartes welcomed the idea of the circulation of the blood, but it appears that his acceptance of it was based on a partial misunderstanding, and he differed from Harvey on the question of the action or function of the heart itself. Most important of all, however, is the fact that the establishment of the circulation of the blood released physiology for a new start in the study of living creatures. Only now could one begin to understand respiration itself properly, or even the digestive and other functions. Given the circulation of the blood running through the arteries and then back by the veins, one could

Chapter 4

The Downfall of Aristotle and Ptolemy

As the crucial stage in the grand controversy concerning the Ptolemaic system does not seem to have been treated organically, and is seldom or never envisaged in its entirety, it is necessary that we should put together a fairly continuous account of it, so that we may survey the transition as a whole. A bird's-eye view of the field should be of some significance for the student of the scientific revolution in general, especially as the battles come in crescendo and rise to their greatest intensity in this part of the campaign.

It would be wrong to imagine that the publication of Copernicus's great work in 1543 either shook the foundations of European thought straight away or sufficed to accomplish anything like a scientific revolution. Almost a hundred and fifty years were needed before there was achieved a satisfactory combination of ideas—a satisfactory system of the universe—which permitted an explanation of the movement of the earth and the other planets, and provided a framework for further scientific development. Short of this, it was only a generation after the death of Copernicus—only towards the close of the sixteenth century—that the period of crucial transition really opened and the conflict even became intense. And when the great perturbations occurred they were the result of very different considerations—the result of events which would have shaken the older cosmos almost as much if Copernicus had never even written his revolutionary work. Indeed, though the influence of Copernicus was as important as people generally imagine it to have been, this influence resulted not so much from the success of his actual system of the skies, but rather from the stimulus which he gave to men who in reality were producing something very different.

When Copernicus's work first appeared it provoked religious objections, especially on Biblical grounds, and since the Protestants were the party particularly inclined to what was called Bibliolatry, some scathing condemnations very soon appeared from their side—for example, from Luther and Melanchthon personally. One may suspect that unconscious prejudice had some part in this, and that the Aristotelian view of the universe had become entangled with Christianity more closely than necessity dictated; for if the Old Testament talked of God establishing the earth fast, the words were capable of elastic interpretation, and Biblical exegesis in previous centuries had managed to get round worse corners than this. In any case, if the Old Testament was not Copernican, it was very far from being Ptolemaic either. And it gives something of a blow to Aristotle and his immaculate fifth essence, surely, when it says that the heavens shall grow old as a garment, and, talking of God, tells us that the stars and the very heavens themselves are not pure in His sight. The prejudice long remained with the Protestants, and when a few years ago the Cambridge History of Science Committee celebrated in the Senate House the tercentenary of the visit to England of the great Czech educator Comenius or Komensky, the numerous orations overlooked the fact that he was anti-Copernican and that his text-books, reprinted in successive editions throughout the seventeenth century, were a powerful influence in the Protestant world on the wrong side of the question. On the other hand, Copernicus was a canon in the Roman Catholic Church and high dignitaries of that Church were associated with the publication of his book. The comparatively mild reception which the new view received on this side led only recently to the enunciation of the view that the Roman Catholics, being slow in the uptake, took nearly fifty years to see that Copernicus was bound to lead to Voltaire. The truth was, however, that the question of the movement of the earth reached the stage of genuine conflict only towards the end of the sixteenth century, as I have said. By that time—and for different reasons altogether—the religious difficulties themselves were beginning to appear more serious than before.

Although Copernicus had not stated that the universe was infinite—and had declared this issue to belong rather to the province of the philosopher—he had been compelled, for a reason which we shall have to consider later, to place the fixed stars at what he called an immeasurable distance away. He was quickly interpreted—particularly by some English followers—as having put the case in favour of an infinite universe; and unless they had some non-religious objections Christians could hardly complain of this, or declare it to be impossible, without detracting from the power and glory of God. Unfortunately, however, that *enfant terrible* amongst sixteenth-century Italian speculators, Giordano Bruno, went further and talked of the actual existence of a plurality of worlds. There arose more seriously than ever before the question: Did the human beings in other worlds need redemption? Were there to be so many appearances of Christ, so many incarnations and so many atonements throughout the length and breadth of this infinite universe? That question was much more embarrassing than the purely Biblical issue which was mentioned earlier; and the unbridled speculations of Bruno, who was burned by the Inquisition for a number of heresies in 1600, were a further factor in the intensification of religious fear on the subject of the Copernican system.

Apart from all this, it is remarkable from how many sides and in how many forms one meets the thesis that is familiar also in the writings of Galileo himself—namely, the assertion that it is absurd to suppose that the whole of this new colossal universe was created by God purely for the sake of men, purely to serve the purposes of the earth. The whole outlay seemed to be too extravagant now that things were seen in their true proportions and the object had come to appear so insignificant. At this later stage the resistance to the Copernican hypothesis was common to both Roman Catholics and Protestants, though in England itself it appears to have been less strong than in most other places. The Protestant astronomer, Kepler, persecuted by the Protestant Faculty at Tübingen, actually took refuge with the Jesuits in 1596. Both the Protestant,

Kepler, and the Roman Catholic, Galileo, ventured into the realms of theology by addressing their co-religionists and attempting to show them that the Copernican system was consistent with a fair interpretation of the words of Scripture. Galileo made excellent use of St. Augustine, and for a time he received more encouragement in the higher ecclesiastical circles in Rome than from his Aristotelian colleagues in the university of Padua. In the long run it was Protestantism which for semi-technical reasons had an elasticity that enabled it to make alliance with the scientific and the rationalist movements, however. That process in its turn greatly altered the character of Protestantism from the closing years of the seventeenth century, and changed it into the more liberalising movement of modern times.

The religious obstruction could hardly have mattered, however, if if had not been supported partly by scientific reasons and partly by the conservatism of the scientists themselves. It has been pointed out by one student that to a certain degree it was the astrologers who were the more ready to be open-minded on this subject in the sixteenth century. Apart from the difficulties that might be involved in the whole new synthesis which Copernicus had provided (and which, as we have seen, included a quasi-superstitious reliance upon the virtues of circles and the behaviour of spheres as such), there were particular physical objections to the attribution of movement to the earth, whether on the plan put forward by Copernicus or in any other conceivable system. Copernicus, as we have seen, had tried to meet the particular objections in detail, but it will easily be understood that his answers, which we have already noted, were not likely to put the matter beyond controversy.

Copernicus himself had been aware that his hypothesis was open to objection in a way that has not hitherto been mentioned. If the earth moved in a colossal orbit around the sun, then the fixed stars ought to show a slight change of position when observed from opposite sides of the orbit. In fact, there is a change but it is so slight that for three centuries after Copernicus it was not detected, and

Copernicus had to explain what then appeared to be a discrepancy by placing the fixed stars so far away that the width of the earth's orbit was only a point in comparison with this distance. If the Ptolemaic theory strained credulity somewhat by making the fixed stars move at so great a pace in their diurnal rotation, Copernicus strained credulity in those days by what seemed a corresponding extravagance—he put the fixed stars at what men thought to be a fabulous distance away. He even robbed his system of some of its economy and its symmetry; for after all the beautiful spacing between the sun and the successive planets he found himself obliged to put a prodigal wilderness of empty space between the outermost planet, Saturn, and the fixed stars. The situation was even more paradoxical than this. When Galileo first used a telescope, one of his initial surprises was to learn that the fixed stars now appeared to be smaller than they had seemed to the naked eye; they showed themselves, he said, as mere pin-points of light. Owing to a kind of blur the fixed stars appear to be bigger than they really ought to appear to the naked eye, and Copernicus, living before that optical illusion had been clarified, was bound to be under certain misapprehensions on this subject. Even before his time some of the fixed stars had seemed unbelievably large when the attempt had been made to calculate their size on the basis of their apparent magnitude. His removal of them to a distance almost immeasurably farther away (while their apparent magnitude remained the same, of course, to the terrestrial observer) made it necessary to regard them as immensely bigger still, and strained a credulity which had been stretched over-far already.

Beyond this there was the famous objection that if the world were rushing from west to east a stone dropped from the top of a tower ought to be left behind, falling therefore well to the west of the tower. The famous Danish astronomer, Tycho Brahe, took this argument seriously, however absurd it might appear to us, and he introduced the new argument that a cannon-ball ought to carry much farther one way than the other, supposing the earth to be in motion. This argument had a novel flavour that made it particularly fashionable in the succeeding period.

In the meantime, however, certain other important things had been happening, and as a result of these it gradually became clear that great changes would have to take place in astronomy—that, indeed, the older theories were unworkable, whether the Copernican hypothesis should happen to be true or not. One of these occurrences was the appearance of a new star in 1572—an event which one historian of science appears to me to be correct in describing as a greater shock to European thought than the publication of the Copernican hypothesis itself. This star is said to have been brighter in the sky than anything except the sun, the moon and Venus—visible even in daylight sometimes—and it shone throughout the whole of the year 1573, only disappearing early in 1574. If it was a new star it contradicted the old view the the sublime heavens knew neither change nor generation or corruption, and people even reminded themselves that God had ceased the work of creation on the seventh day. Attempts were made to show that the star existed only in the sublunary region, and even Galileo later thought it necessary to expose the inaccurate observations which were selected from the mass of available data to support this view. After all, Copernicus had only put forward an alternative theory of the skies which he claimed to be superior to the ancient one. Now, however, men were meeting inconvenient facts which sooner or later they would have to stop denying.

In 1577 a new comet appeared, and even some people who disbelieved the Copernican theory had to admit that it belonged to the upper skies, not to the sublunary regions—the more accurate observations which were now being made had altered the situation in regard to the observation of the whereabouts of comets. As this one cut a path straight through what were supposed to be the impenetrable crystal spheres that formed the skies, it encouraged the view that the spheres did not actually exist as part of the machinery of the heavens; Tycho Brahe, conservative though he was in other respects, henceforward declared his disbelief in the reality of these orbs. In the last quarter of the sixteenth century Giordano Bruno, whom I have already mentioned, pictured the planets and stars floating in empty space, though it now became more difficult than ever to say why they moved and how they were kept in their regular paths. Also the Aristotelian theory that comets were formed out of mere exhalations from the earth, which ignited in the sphere of fire—all within the sublunary realm—was no longer tenable. And those who did not wish to fly in the face of actual evidence began to modify the Aristotelian theory in detail—one man would say that the upper heavens were not unchangeable and uncorruptible; another would say that the very atmosphere extended throughout the upper skies, enabling the exhalations from the earth to rise and ignite even in the regions far above the moon. Quite apart from any attack which Copernicus had made upon the system, the foundations of the Ptolemaic universe were beginning to shake.

It is particularly towards the end of the sixteenth century that we can recognize the extraordinary intermediate situation which existed—we can see the people themselves already becoming conscious of the transitional stage which astronomical science had reached. In 1589 one writer, Magini, said that there was a great demand for a new hypothesis which would supersede the Ptolemaic one and yet not be so absurd as the Copernican. Another writer, Maestlin, said that better observations were needed than either those of Ptolemy or those of Copernicus, and that the time had come for "the radical renovation of astronomy." People even put forward the view that one should drop all hypotheses and set out simply to assemble a collection of more accurate observations. Tycho Brahe replied to this that it was impossible to sit down just to observe without the guidance of any hypothesis at all.

Yet that radical renovation of astronomy which Maestlin required was being carried out precisely in the closing years of the sixteenth century; and Tycho Brahe was its first leader, becoming important not for his hypotheses but precisely because of what has been called the "chaos" of observations that he left behind for his successors. We have seen that in the last quarter of the sixteenth century he achieved practically all that in fact was achieved, if not all that was possible, in the way of pre-telescopic observation. He greatly improved the instruments and the accuracy of observation. He followed the planets throughout the whole of their courses, instead of merely trying to pick them out at special points in their orbits. We have noticed also his anti-Copernican fervour, and in one respect his actual systematising was important, though his theories were not justified by events; and when he had made his observations he did not follow them up with any development of them since he was not a remarkable mathematician. He attempted, however, to establish a compromise between the Ptolemaic and the Copernican systems—some of the planets moving around the sun, but then the sun and its planetary system moving in a great sweep around the motionless earth. This is a further illustration of the intermediate and transitional character of this period, for his compromise gained a certain following; he complained later that other men pretended to be the inventors of it; and after a certain period in the seventeenth century this system secured the adhesion of those who still refused to believe in the actual movement of the earth. He was not quite so original as he imagined, and his compromise system has a history which goes back to much earlier times.

Still more significant was the fact that the chaos of data collected and recorded by Tycho Brahe came into the hands of a man who had been his assistant for a time, Johann Kepler, the pupil of the very person, Maestlin, who had demanded a renovation of astronomy. Kepler, therefore, emerges not merely as an isolated genius, but as a product of that whole movement of renovation which was taking place at the end of the sixteenth century. He had the advantage over Tycho Brahe in that he was a great mathematician, and he could profit from considerable advances that had taken place in mathematics during the sixteenth century. There was one further factor which curiously assisted that renovation of astronomy which we are examining at the moment, and it was a factor of special importance if the world was to get rid of the crystal spheres and see the planets merely floating in empty space. An Englishman, William Gilbert, published a famous book on the magnet in 1600 and laid himself open to the gibes of Sir Francis Bacon for being one of those people so taken up with their pet subject of research that they could only see the whole universe transposed into the terms of it. Having made a spherical magnet called a *terrella,* and having found that it revolved when placed in a magnetic field, he decided that the whole earth was a magnet, that gravity was a form of magnetic attraction, and that the principles of the magnet accounted for the workings of the Copernican system as a whole. Kepler and Galileo were both influenced by this view, and with Kepler it became an integral part of his system, a basis for doctrine of almost universal gravitation. William Gilbert provided intermediate assistance therefore—brought a gleam of light—when the Aristotelian cosmos was breaking down and the heavenly bodies would otherwise have been left drifting blindly in empty space.

With all these developments behind him, therefore, the famous Kepler in the first thirty years of the seventeenth century "reduced to order the chaos of data" left by Tycho Brahe, and added to them just the thing that was needed—mathematical genius. Like Copernicus he created another world-system which, since it did not ultimately prevail, merely remains as a strange monument of colossal intellectual power working on insufficient materials; and even more than Copernicus he was driven by a mystical semi-religious fervour—a passion to uncover the magic of mere numbers and to demonstrate the music of the spheres. In his attempt to disclose mathematical sympathies in the machinery of the skies he tried at one moment to relate the planetary orbits to geometrical figures, and at another moment to make them correspond to musical notes. He was like the child who having picked a mass of wild flowers tries to arrange them into a posy this way, and then tries another way, exploring the possible combinations and harmonies. He has to his credit a collection of discoveries and conclusions—some of them more ingenious than useful—from which we today can pick out three that have a permanent importance in the history of astronomy. Having discovered in the first place that the planets did not move at a uniform speed, he set out to find order somewhere, and came upon the law that if a line were drawn from a given planet to the sun that line would describe equal areas in equal times. At two different points in his calculations it would appear that he made mistakes, but the conclusion was happy, for the two errors had the effect of cancelling one another out. Kepler realised that the pace of the planet was affected by its nearness to the sun—a point which encouraged him in his view that the planets were moved by a power actually emitted by the sun.

His achievements would have been impossible without that tremendous improvement in observation which had taken place since the time of Copernicus. He left behind him great masses of papers which help the historian of science to realise better than in the case of his predecessors his actual manner of work and the stages by which he made his discoveries. It was when working on the data left by Tycho Brahe on the subject of the movements of Mars that he found himself faced with the problem of accounting for the extraordinary anomalies in the apparent orbit of this planet. We know how with colossal expenditure of energy he tried one hypothesis after another, and threw them away, until he reached a point where he had a vague knowledge of the shape required, decided that for purposes of calculation an ellipse might give him at any rate approximate results, and then found that an ellipse was right—a conclusion which he assumed then to be true also for the other planets.

Some people have said that Kepler emancipated the world from the myth of circular motion, but this is hardly true, for from the time of the ancient Ptolemy men had realised that the planets themselves did not move in regular circles. Copernicus had been aware that certain combinations of circular motion would provide an elliptical course, and even after Kepler we find people accounting for the new elliptical path of the planets by reference to a mixture of circular movements. The obsession on the subject of circular motion was disappearing at this time, however, for other reasons, and chiefly because the existence of the hard crystal spheres was ceasing to be credible. It had been the spheres, the various inner wheels of the vast celestial machine, that had enjoyed the happiness of circular motion, while the planet, recording the resultant effect of various compound movements, had been realised all the time to be pursuing a more irregular course. It was the circular motion of the spheres themselves that symbolised the perfection of the skies, while the planet was like the rear lamp of a bicycle—it might be the only thing that could actually be seen from the earth, and it dodged about in an irregular manner; but just as we know that it is really the man on the bicycle who matters, though we see nothing save the red light, so the celestial orbs had formed the essential machinery of the skies, though only the planet that rode on their shoulder was actually visible. Once the crystal spheres were eliminated, the circular motion ceased to be the thing that really mattered—henceforward it was the actual path of the planet itself that fixed one's attention. It was as though the man on the bicycle had been proved not to exist, and the rear lamp, the red light, was discovered to be sailing on its own account in empty space. The world might be rid of the myth of circular motion, but it was faced with more difficult problems than ever with these lamps let loose and no bicyle to attach them to. If the skies were like this, men had to discover why they remained in any order at all—why the universe was not shattered by the senseless onrush and the uncontrollable collidings of countless billiard-balls.

Kepler believed in order and in the harmony of numbers, and it was in his attempt to fasten upon the music of the spheres that he discovered, amongst many other things, that third of his series of planetary laws which was to prove both useful and permanent—namely, the law that the squares of the period of the orbit were proportional to the cubes of their mean distances from the sun. By this time Kepler was mystical in a sense somewhat different from before—he was no longer looking for an actual music of the spheres which could be heard by God or man, or which should be loaded with mystical content. The music of the spheres was now nothing more or less to him than mathematics as such—the purely mathematical sympathies that the universe exhibited—so that what concerned him was merely to drive ahead, for ever eliciting mathematical proportions in the heavens. In fact, we may say that this worship of numerical patterns, of mathematical relations as such, took the place of the older attempt, that was still visible in Galileo, to transpose the skies into terms of circles and spheres, and became the foundation of a new kind of astronomy. It is in this particular sense that Kepler can most properly be described as having provided an improvement upon the old superstition which had hankered only after circular motion. Furthermore, by the same route, Kepler became the apostle of a mechanistic system—the first one of the seventeenth-century kind—realising that he was aspiring to turn the universe into pure clockwork, and believing that this was the highest thing he could do to glorify God. It will be necessary to glance at the Keplerian system as a whole when we come to the problem of gravitation at a later stage of the story. We must note that, of course, Kepler believed in the motion of the earth, and showed that if this supposition were accepted the movement conformed to the laws which he had discovered for the planets in general.

Besides Kepler's three planetary laws, one final addition was being made in this period to the collection of material that spelt the doom of Ptolemy and Aristotle. In 1609 Galileo, having heard of the discovery of the telescope in Holland, created a telescope for himself, though not before an actual sample of the Dutch instrument had appeared in Venice. Instantly the sky was filled with new things and the conservative view of the heavenly bodies became more completely untenable than ever. Two items were of particular importance. First, the discovery of the satellites of Jupiter provided a picture of what might be described as a sort of miniature solar system in itself. Those who had argued that the moon obviously goes round the earth, *ergo* in a regular heaven the celestial bodies must move about the same centre, were now confronted with the fact that Jupiter had its own moons, which revolved around it, while both Jupiter and its attendants certainly moved together either around the sun as the Copernicans said, or around the earth according to the system of Ptolemy. Something besides the earth could be shown to operate therefore as the centre of motions taking place in the sky. Secondly, the sunspots now became visible and if Galileo's observations of them were correct they destroyed the basis for the view that the heavens were immaculate and unchanging. Galileo set out to demonstrate that the spots were, so to speak, part of the sun, actually revolving with it, though the Aristotelians tried to argue that they were an intervening cloud, and that some of Galileo's discoveries were really the result of flaws in the lenses of his telescope. Galileo was seriously provoked by these taunts and at this point of the story of the whole controversy with the Aristotelians flared up to an unprecendented degree of intensity, not only because the situation was ripe for it, but because Galileo, goaded to scorn by university colleagues and monks, turned his attention from questions of mechanics to the larger problem of the Aristotelian issue in general. He ranged over the whole field of that controversy, bringing to it an amazing polemical imagination, which goaded the enemy in turn.

His intervention was particularly important because the point had been reached at which there was bound to be a complete impasse unless the new astronomy could be married somehow to the new science of dynamics. The Aristotelian cosmos might be jeopardised, and indeed was doomed to destruction by the recent astronomical disclosures; yet these facts did not in the least help the enquirers over the original hurdle—did not show them how to square the movement of the earth itself with the principles of Aristotelian mechanics or how to account for the motions in the sky. Copernicus had taken one course in treating the earth as virtually a celestial body in the Aristotelian sense—a perfect sphere governed by the laws which operated in the higher reaches of the skies. Galileo complemented this by taking now the opposite course—rather treating the heavenly bodies as terrestrial ones, regarding the planets as subject to the very laws which applied to balls sliding down inclined planes. There was something in all this which tended to the reduction of the whole universe to uniform physical laws, and it is clear that the world was coming to be more ready to admit such a view.

After his construction of a telescope in 1609 and the disturbing phenomena which it quickly revealed in the skies, Galileo's relations with the Peripatetics—the worshippers of Aristotle—at the university of Padua became intensely bitter. Though for a time he met with support and encouragement in high places and even in Rome itself, the intensified controversy led to the condemnation of the Copernican hypothesis by the Congregation of the Index in 1616. This did not prevent Galileo from producing in the years 1625-29 the series of Dialogues on *The Two Principal World-Systems* which he designed to stand as his *magnum opus* and which were to lead to his condemnation. This book traversed the whole range of anti-Aristotelian argument, not merely in the realm of astronomy, but in the field of mechanics, as though seeking to codify the entire case against the adherents of the ancient system. It stands as a testimony to the fact that it was vain to attack the Aristotelian teaching merely at a single point—vain to attempt in one corner of the field to reinterpret motion by the theory of impetus as the Parisian scholastics had done—which was only like filling the gap in one jigsaw puzzle with a piece out of a different jigsaw puzzle altogether. What was needed was a large-scale change of design—the substitution of one highly dovetailed system for another—and in a sense it appeared to be the case that the whole Aristotelian synthesis had to be overturned at once. And that is why Galileo is so important; for, at the strategic moment, he took the lead in a policy of simultaneous attack on the whole front.

The work in question was written in Italian and addressed to a public somewhat wider than the realm of learning—wider than that university world which Galileo had set out to attack. Its argument was conducted much more in the language of ordinary conversation, much more in terms of general discourse, than the present-day reader would expect—the *Dialogues* themselves are remarkable for their literary skill and polemical scorn. Galileo paid little attention to Kepler's astronomical discoveries—remaining more Copernican in his general views, more content to discuss purely circular motion in the skies, than the modern reader would expect to be the case. He has been regarded as unfair because he talked only of two principal world-systems, those of Ptolemy and Copernicus, leaving the new systems of Tycho Brahe and Johann Kepler entirely out of account. In his mechanics he was a little less original than most people imagine, since, apart from the older teachers of the impetus-theory, he had had more immediate precursors, who had begun to develop the more modern views concerning the flight of projectiles, the law of inertia and the behaviour of falling bodies. He was not original when he showed that clouds and air and everything on the earth—including falling bodies—naturally moved round with the rotating earth, as part of the same mechanical system, and in their relations with one another were unaffected by the movement, so that like the objects in the cabin of a moving ship, they might appear motionless to anybody moving with them. His system of mechanics did not quite come out clear and clean, did not even quite explicitly reach the modern law of inertia, since here again he had not quite disentangled himself from obsessions concerning circular motion. It was chiefly in his machanics, however, that Galileo made his contributions to the solution of the problem of the skies; and here he came so near to the mark that his successors had only to continue their work on the same lines—future students were able to read back into his writings views which in fact were only put forward later. Galileo's kind of mechanics had a strategic place in the story, for they had to be married to the astronomy of Kepler before the new scientific order was established. And the new dynamics themselves could not be developed merely out of a study of terrestrial motion. Galileo is important because he began to develop them with reference to the behaviour of the heavenly bodies too.

At the end of everything Galileo failed to clinch his argument—he did not exactly prove the rotation of the earth—and in the resulting situation a reader could either adopt his whole way of looking at things or could reject it *in toto*—it was a question of taking over that whole realm of thought into which he had transposed the question. It was true that the genuinely scientific mind could hardly resist the case as a whole, or refuse to enter into the new way of envisaging the matter; but when Galileo's mouthpiece was charged in the *Dialogues* with having failed to prove his case—having done nothing more than explain away the ideas that made the movement of the earth seem impossible—he seemed prepared to admit that he had not demonstrated the actual movement, and at the end of Book III he brought out his secret weapon—he declared that he had an argument which would really clinch the matter. We know that Galileo attached a crucial importance to this argument, which appears in the fourth book, and, in fact, he thought of taking the title of the whole work from this particular part of it. His argument was that the tides demonstrated the movement of the earth. He made a long examination of them and said that they were caused, so to speak, by the shaking of the vessel which contained them. This seemed to contradict his former argument that everything on the earth moved with the earth, and was as unaffected by the movement as the candle in the cabin of a moving ship. It was the combination of motions, however—the daily rotation together with the annual movement, and the accompanying strains and changes of pace—which produced the jerks, he said, and therefore set the tides in motion. Nothing can better show the transitional stage of the questions even now than the fact the Galileo's capital proof of the motion of the earth was a great mistake and did nothing to bring the solution of the question nearer.

Aristotelian physics were clearly breaking down, and the Ptolemaic system was split from top to bottom. But not till the time of Newton did the satisfactory alternative system appear; and though the more modern of the scientists tended to believe in the movement of the earth from this time, the general tendency from about 1630 seems to have been to adopt the compromise system of Tycho Brahe. In 1672 a writer could say that the student of the heavens had four different world-systems from which to choose, and there were men who even talked of seven. Even at this later date an enquirer could still come forward—as Galileo had done—and claim that at last he had discovered the capital argument. The long existence of this dubious, intermediate situation brings the importance of Sir Isaac Newton into still stronger relief. We can better understand also, if we cannot condone, the treatment which Galileo had to suffer from the Church for a presumption which in his dialogues on *The Two Principal World-Systems* he had certainly displayed in more ways than one.

Although Galileo's most famous writings appeared in the 1630's they were the fruits of work which had been done at an earlier date. The second quarter of the seventeenth century represents really a new generation—that of the disciples of Galileo, and particularly of those who followed him in his capacity as the founder of modern mechanics. In the 1630's and 1640's his arguments are carried to a further stage, and the essential theme of the story assembles itself around a group of interrelated workers whose centre seems to be Paris, though there are connections also with Holland and Italy.

The group in question includes Isaac Beeckman (1588-1637) in Holland, a man who stimulated others to take an interest in important problems and initiated a number of ideas. Next to him comes Marin Mersenne (1588-1648), not himself a great discoverer, but a central depot of information and a general channel of communication—a man who provoked enquiries, collected results, set one scientist against another, and incited his colleagues to controversy. Next again in point of age comes Pierre Gassendi (1592-1655), a philosopher and writer of scientific biographies, who possessed an encyclopædic knowledge of the sciences of the time. And after him comes René Descartes (1596-1650), though in many respects he stands out as a lonely worker, mathematician, physicist and philosopher all in one. Gilles de Roberval (1602-75) comes next, and he is an original figure, essentially a mathematician; while Galileo's pupil, Evangelista Torricelli (1608-47), also comes into the picture on occasion. Even the famous Pascal, and also Christian Huygens, were brought into contact with the circle as young men in the latter part of the period, and helped to form the bridge to a later generation, their fathers having been connected with the group. The Englishman, Thomas Hobbes, first began to develop his views on the physical universe after he had made contact with Mersenne and his friends.

These are the people who carry the argument a stage further. Though they are disciples of Galileo in the field of mechanics they are at first inclined to adopt a cautious attitude in respect of his cosmology. Some of them feel that Galileo has not established the case for Copernicus, though they may like the Copernican system because it is more economical and more æsthetic than the ancient one. What they have chiefly taken over is Galileo's way of mathematising a problem, and what affects them most of all perhaps is the establishment of the modern principle of inertia—the thesis that things will continue their movement in a straight line until something actually intervenes to check or alter their motion. This principle is important because it provides the starting-point for a new science of dynamics.

Work was continued on the problem of falling bodies and on questions of hydrostatics; and the atmosphere itself was now being examined on mechanical principles. Round about 1630, in various regions independently, work was conducted on the assumption that the air has weight; the problem of the possible existence of a vacuum became alive again; and there occurred those developments which carry us from Galileo to Torricelli's famous experiment—to the barometer and the air-pump. The old way of explaining things through the assumed existence of secret sympathies in various forms of matter, or through nature's "abhorrence" of a vacuum, was derided—only mechanical explanations would serve. The magnet was still a serious problem, because it seemed to confirm the idea of sympathetic attraction; but there was now a tendency to believe that it would some day be explicable on mechanical principles. One was now less inclined to believe in the ability of the magnet to recognise an adulterous woman or to bring about peace between man and wife.

The war on Aristotle still continued; and this meant also war against medieval scholasticism and against the modern conservative followers of Aristotle, the Peripatetics, who retained their place in the universities even after this period was over. But the warfare was equally against the so-called naturalism of the Renaissance—the belief in pan-psychism and animism, which gave everything a soul and saw miracles everywhere in nature. It was partly in the name of religion itself that Renaissance naturalism was attacked, and the Christians helped the cause of modern rationalism by their jealous determination to sweep out of the world all miracles and magic except their own. Some of this new generation of scientists argued that Christian miracles themselves could not be vindicated unless it could be assumed that the normal workings of the universe were regular and subject to law. In the circle around Mersenne in the 1630's the idea of a complete mechanistic interpretation of the universe came out into the open, and its chief exponents were the most religious men in the group that we are discussing. They were anxious to prove the adequacy and the perfection of Creation—anxious to vindicate God's rationality.

The advent of the printed book on the one hand, and of the woodcut and the engraving on the other, had greatly transformed the problem of scientific communication from the time of the Renaissance, though even in the sixteenth century it is surprising to see how local the effect of original work might be. Before the close of the century the correspondence between scientific workers became significant—particularly perhaps that between the astronomers, who found it valuable to compare the observations that were made in one place and another. From the time of Galileo the development of modern science appears much more as a general movement; it is much less capable of reconstruction as a case of isolated endeavour. The experimental method became quite the fashion amongst groups of people both inside the universities and outside them; and men who had previously cultivated antiquity or collected coins began to regard it as a mark of culture to patronise science and experiments too, and to collect rare plants or curiosities in nature. Amongst the clergy and the university teachers, the doctors and the gentry, there would emerge enthusiastic amateurs, some of them attracted by the love of marvels, by mechanical tricks and toys, or by the fantastic side of nature. Indeed, a good number of the famous names of the seventeenth century would seem to have belonged to this class.

To a certain degree the scientist made use of the forms of communication which already existed for other purposes in that period; and the antecedents of the scientific societies are the literary clubs of the sixteenth century and the groups of people who assembled to discuss philosophy at the time of the Renaissance. It was customary for people to meet in informal societies and read news-letters which had been written from correspondents abroad—letters which would describe not only political events but recent publications and movements of ideas. Scientific works and even experiments would come to be included in the readings and discussions. In some cases, those who were interested in science would feel that the news was too political, and would try to make the proceedings more scientific or would be inclined to break away and form a purely scientific circle of their own. One group at the house of the French historian De Thou had been composed of scholars, men of letters and members of the professions, and later, for a number of decades, it sat under the brothers Dupuy, serving as a bureau for the exchange of foreign news and being at times largely political in its interest. It did not always move on the same course, however, and we find members of the scientific movement—Mersenne and Gassendi, for example—

amongst those who attended its gatherings. Henry Oldenburg, later Secretary of the Royal Society, went to the meetings in 1659-60. Between 1633 and 1642 weekly conferences were held at the house of Théophraste Renaudot in Paris, and here a weekly pamphlet was published. They were called Conférences de Bureau d'Adresse, and the discussions dealt with concepts such as those of First Matter and Cause, subjects like air, water, atoms, dew and fire, mythical creatures like the unicorn and phoenix, but also novels, dancing, the education of women and the status of trade. These conferences had an influence in England in the late 1640's.

More serious groups or societies or academies existed amongst actual scientific workers from the very beginning of the seventeenth century, however; and here the priority seems to belong to a circle in Rome called the Accademie dei Lincei, which ran from 1600 to 1657, with an interval before 1609 when they were broken up because they were suspected of poisonings and incantations. They met at the house of their patron, a Duke, but hoped to establish their own museum, library, laboratory, botanical garden and printing office, as well as to organise subordinate branches in various parts of the world. From 1609 their written proceedings form the earliest recorded publication of a scientific society. Galileo himself was an active member, and he made a microscope for the society, which published one or two of his important works.

Similar historical significance attaches to the conferences founded by Mersenne in 1635 and kept up by him until his death in 1648, conferences which brought together with more or less regularity the mathematicians and physicists, Gassendi, Desargues, Roberval, Descartes, the elder and the younger Pascal, and many others. Mersenne, we are told, is responsible more than any other single person for the establishment of the intellectual centre of Europe in Paris during the middle third of the seventeenth century. He himself conducted a universal correspondence, passed problems from one scientist to

Chapter 5

The Experimental Method in the Seventeenth Century

IT IS NOT always realised to what a degree the sciences in the middle ages were a matter for what we today would describe as literary transmission, and came into European history as a heritage from ancient Greece and imperial Rome. Nobody can examine the actual state of scientific knowledge in, say, the tenth century A.D. without realising what had been lost both in scholarship and in technique—indeed, in civilisation as a whole—since the days of ancient Athens and ancient Alexandria, or even since the time when St. Augustine flourished. Nobody who has any picture of Europe as it emerged from the dark ages, or any impression of our Anglo-Saxon forefathers one or two centuries before the Norman Conquest, will imagine that the world was then in a condition to discover by its own enquiries and experiments the scientific knowledge which Athens and Alexandria had attained at a time when their civilisation was at its peak. Actual contact with the science of the ancient world had to be re-established by the unearthing of texts and manuscripts, or by the acquisition of translations and commentaries from peoples like the Arabs or the subjects of the Byzantine Empire, who already possessed, or had never lost, the contact. That process of recovery reached its climax and came to full consciousness in the period of what we call the Renaissance. It would have taken many hundreds of years more if the middle ages had had, so to speak, to find the same things out for themselves—to re-create so much of the development of science by independent enquiry and unaided research.

All this helps to explain why so much of the history of medieval thought rests on a framework of dates which are really dates in the literary transmission of ancient science and scholarship. Historians find it of primary importance to discover at what date such and such a work of Aristotle was resuscitated in western Europe; or when this or that scientific treatise became available through an Arabian translation, and—better still—when western Europe was able to acquire the authentic text in the original Greek. The process was not stopped by any reluctance on the part of Catholic Europe to learn from the infidel Arabians or the Byzantine schismatics or even the pagan Greeks. Nor is it known that there was any opportunity which the middle ages missed—any great store of science that they turned their backs upon because it was tainted with paganism or infidelity. Because the intelligentsia in the middle ages was a clerical one and the intellectual leadership was religious in character, such natural science as existed was the more likely to keep the subordinate place it had always had in a larger philosophical system—what we call "natural scientists" could hardly be said to have existed then, in any significant sense of the term. Because the purely literary transmission was so important, that thing which we call science, and which might rather be called natural philosophy, was first and foremost a series of ancient texts upon which one commentary after another would be compiled, often by people writing, so to speak, at a desk. If even at the Renaissance philology was considered the queen of the sciences, this was because the man who was master of the classical languages did in fact hold the key position. We can still read the letters of humanists who cursed their fate because they had to ruin their style by translating works of physics from the Greek.

So in the middle ages men found themselves endowed with an explanation of the physical universe and the workings of nature which had fallen upon them out of the blue, and which they had taken over full-grown and ready-made. And they were infinitely more the slaves of that intellectual system than if they had actually invented it themselves, developing it out of their own original researches and their own wrestlings with truth. There even seems to have been a perceptible hurdle here and there where there was a gap in the transmission—where patches of ancient scholarship had still remained undiscovered. We have already noticed, for example, certain tendencies in fourteenth-century Paris which are considered to have been nipped in the bud because of a deficiency in mathematics—a deficiency somewhat rectified by a further recovery of ancient texts in the period of the Renaissance. Under such conditions the chief openings for independent thought—the chief controversies in the sixteenth century even—occurred at those places where the ancient writers were found to have differed from one another. And though in the later middle ages there were men who were doing experiments and pushing back the frontiers of thought, they were, for the most part, like the theorists of the impetus, only playing on the margin of that Aristotelian system which in the year 1500 must have appeared at least as valid to a rational thinker as it could have done fifteen hundred years before. Though there were men in the later middle ages who were carefully observing nature, and improving greatly in the accuracy of their observations, these tended to compile encyclopædias of purely descriptive matter. When there was anything that needed to be explained these men would not elicit their theories from the observations themselves—they would still draw on that whole system of explanation which had been provided for them by the ancient philosophy. Sir Francis Bacon, early in the seventeenth century, complained of this divorce between observation and explanation, and it was part of his purpose to show how the latter ought to arise out of the former.

So far as one can see, the mathematics of ancient Alexandria, acquired at the time of the Renaissance, and the works of Archimedes, made generally available in translation in 1543, represent the last pocket of the science of antiquity which was recovered in time to be an ingredient or a factor in the formation of our modern science. As we have already seen, this was a body of knowledge which, so far as one can judge, it was necessary to recover before all the components of the scientific movement could be assembled together and the autonomous efforts of scientific enquirers—of a new crowd of pioneers in research—could properly be put into gear. And it is remarkable how quickly things began to move, once all the ingredients, so to speak, had at least been collected together. Early in the seventeenth century, as we have already seen, the ancient explanation of the universe—the framework of existing science—was palpably breaking down. There was beginning to emerge what contemporaries clearly recognised as a scientific revolution, and what to us is the dawn of modern science.

Now, if we are seeking to understand this birth of modern science we must not imagine that everything is explained by the resort to an experimental mode of procedure, or even that experiments were any great novelty. It was commonly argued, even by the enemies of the Aristotelian system, that that system itself could never have been founded except on the footing of observations and experiments—a reminder necessary perhaps in the case of those university teachers of the sixteenth and seventeenth centuries who still clung to the old routine and went on commentating too much (in what we might call a "literary" manner) upon the works of the ancient writers. We may be surprised to note, however, that in one of the dialogues of Galileo, it is Simplicius, the spokesman of the Aristotelians—the butt of the whole piece—who defends the experimental method of Aristotle against what is described as the mathematical method of Galileo. And elsewhere it is the man speaking as the mouthpiece of Galileo himself who says that though Aristotle only gives reasoning to prove that such and such a thing must be the case, still this is only Aristotle's way of demonstrating the thesis—the actual discovery of it must have been the result of experiment. We have already seen how the medical students and the medical university of Padua were ahead of most other people in their regard for experiment, and the most remarkable result of the experimental method that we have met with so far in these lectures is William Harvey's treatise on the circulation of the blood. Yet it was not in the biological sciences that the Aristotelian way of attacking the problem was to receive its spectacular overthrow. It was not there that the scientific revolution found its centre or its pivot—on the contrary, we shall have to study later the effects of the scientific revolution as they come by reflection, so to speak (and at a second remove), upon the biological and other sciences. What is more remarkable still is the fact that the science in which experiment reigned supreme—the science which was centred in laboratories even before the beginning of modern times—was remarkably slow, if not the slowest of all, in reaching its modern form. It was long before alchemy became chemistry, and chemistry itself became in the full sense of the word quantitative in its method, instead of being qualitative, after the manner of ancient science.

It may be interesting in this connection to glance at what perhaps is the most famous experiment of the scientific revolution—what an historian of science declared in 1923 to be "one of the outstanding achievements of scientific history." It comes from the vague story of a disciple and a somewhat romantic biographer of Galileo, who said that his teacher had dropped two bodies of different weights from the tower of Pisa to prove that Aristotle was wrong in his view that they would fall at paces proportional to their weights. Later historians of science filled in the details, so that in a work published in 1918 the final precision was attained, and we learn how this martyr of science climbed the leaning tower of Pisa with a one-hundred-pound cannon ball under one arm and a one-pound ball under the other. Even Dr. Singer repeated the story in 1941 in his history of science, where he calls it "the most famous of experiments" and attributes it to the year 1591. None of the vast crowd who are supposed to have observed the experiment gave any evidence on its behalf—though, as we shall see, there was a particular reason why they should have done so if they had actually been witnesses—and the writings of Galileo give no confirmation of the story. On the contrary, the writings of Galileo showed that he had tried the experiment several times in his youth with the opposite result—he said in one of his juvenile works that he had tested the matter on many occasions from a high tower and that in his experience a lump of lead would very soon leave a lump of wood behind. The supposed experiment had actually been tried by another scientist, Simon Stevin of Bruges, and was recorded in a book published in 1605. Stevin, however, dropped balls of lead only from a height of thirty feet, and, considering how little was known in those days about the effects of such things as air-resistance, the Aristotelians were perhaps not unreasonable in saying that the result was not conclusive—you needed to try the experiment from a great height.

Galileo, who in his youth indulged in curious speculations concerning the behaviour of falling bodies, ought to have been in a position to appreciate that argument; for, again in one of his early works, he had even insisted that it was useless to drop bodies from the top of a tower—the height would need to be doubled before it was possible to form a proper judgment, he said. To crown the comedy, it was an Aristotelian, Coresio, who in 1612 claimed that previous experiments had been carried out from too low an altitude. In a work published in that year he described how he had improved on all previous attempts—he had not merely dropped bodies from a high window, he had gone to the very top of the tower of Pisa. The larger body had fallen more quickly than the smaller one on this occasion, and the experiment, he claimed, had proved Aristotle to have been right all the time. Coresio's work was published in Florence, and it does not appear that either Galileo or anybody else challenged the truth of the assertion, though the date is long after that of the alleged incident in the life of Galileo.

In reality, the predecessors of Galileo had for some time been gradually approaching the settlement of the problem on different lines altogether. At first they had moved timidly and had argued that different weights of the same substance would fall simultaneously; though there might be a difference in pace, they said, if the comparisons were between different substances altogether. Galileo, in fact, uses the argument employed by his predecessors—they had reasoned that two tiles each weighing a pound and dropped at the same moment would fall to the ground at precisely the same time. Fastened together, end to end, they would still descend at the pace at which they had fallen when dropped merely side by side. And if one were fastened on the top of the other, still it would not press down more heavily than before, and therefore it would do nothing to press its lower partner to fall any more quickly either. In other words, the predecessors of Galileo had reasoned their way to the answer to this particular problem, and neither they nor Galileo showed any willingness to alter the conclusion merely because the experimental method had failed to confirm their judgment. In his youth Galileo had held the view for a time that falling bodies did not accelerate—at least, they only accelerated at the beginning of their fall, he said, until they got into proper going form. Even on this point he was not to be put off by mere observation. It was in this connection that he refused to be deterred by the results of an experiment made from a tower, and said that it would be necessary to drop things from twice that height before the experiment could be regarded as decisive. As an appendix to the whole story I may note the existence of a controversy on the question whether Aristotle himself held the views for which this crucial experiment was supposed to have brought him into discredit. The matter is irrelevant, however, as at any rate the Aristotelians of the seventeenth century held these views and accepted the issue as a fair one.

In connection with this and many similar problems, it would be somewhere near the truth if one were to say that for about fifty years there has been considerable comment on what are called the "thought-experiments" of Galileo. In some of his works one can hardly fail to notice the way in which he would assert: "If you were to do this thing, then this other particular thing *would* happen"; and on some occasions it would appear to be the case that he was wrong in his inference—on some occasions nobody stops to worry if one of the parties in the dialogues even makes the point that the experiment has never been tried. It is curious also how often Galileo makes use of these "thought-experiments" in regard to those points of mechanics that affect the question of the rotation of the earth

—how often he resorts to them when he is meeting the arguments that were the chief stock-in-trade of the Aristotelians. He discusses what would happen if you were to drop a stone from the top of the mast of a ship (*a*) when the vessel was moving and (*b*) when the vessel was at rest. Much later, in 1641, a considerable sensation was caused by Gassendi, who actually tried the experiment and published the result, which on this occasion confirmed the thesis of Galileo. There was in France a younger contemporary and admirer of Galileo, called Mersenne, who, though a disciple of the great Italian in mechanics, was unable to feel convinced by the arguments which had been put forward in favour of the rotation of the earth. He came across Galileo's "thought-experiments" in this field and on one occasion after another we find him making the significant comment: "Yes, only the experiment has never been tried." As, later, he began to show himself more sympathetic to the Copernican point of view, Mersenne revealed that even now it was a different form of reasoning that appealed to him—a type of argument belonging to a period long before the time of Galileo. He said: "If I could be convinced that God always did things in the shortest and easiest way, then I should certainly have to recognise the fact that the world does move."

The scientific revolution is most significant, and its achievements are the most remarkable, in the fields of astronomy and mechanics. In the former realm the use of experiment in any ordinary sense of the word can hardly be expected to have had any relevance. In regard to the latter we may recall what we observed when we were dealing with the problem of motion—how it seemed reasonable to say that the great achievement was due to a transportation taking place in the mind of the enquirer himself. Here was a problem which only became manageable when in a certain sense it had been "geometrised," so that motion had come to be envisaged as occurring in the emptiness of Archimedean space. Indeed, the modern law of inertia—the modern picture of bodies continuing their motion in a straight line and away to infinity—was hardly a thing which the human mind would ever reach by an experiment, or by any attempt to make observation more photographic, in any case. It depended on the trick of seeing a purely geometrical body sailing off into a kind of space which was empty and neutral—utterly indifferent to what was happening—like a blank sheet of paper, equally passive whether we draw on it a vertical or a horizontal line.

In the case of the Aristotelian system the situation had been different—it had always been impossible to forget that certain parts of the universe had a special "pull." There were certain directions which it was fundamental to regard as privileged directions. All lines tended to be attracted to the centre of the earth. Under this system it was not possible to make the required abstraction, and, for example, to draw a simple straight line to represent a body flying off at a tangent—flying off with determination and rectitude into infinite space. It was necessary that the line should curl round to the bottom of the paper, for the very universe was pulling it down, dragging the body all the time towards the centre of the earth. At this point even Galileo was imperfect. He did not attain the full conception of utterly empty, utterly directionless, Euclidean space. That is why he failed to achieve the perfect formulation of the modern law of inertia, for he believed that the law of inertia applied to motion in a circle; and here he was wrong—what we call "inertial motion" must be movement along a straight line. When he talked of a perfectly spherical ball riding off to infinity on a perfectly smooth horizontal plane, he showed his limitations; for he regarded the horizontal plane as being equidistant from the centre of the earth, and pictured it as a plane that actually went round the earth; so that he could seize upon even this as a form of circular motion. And, though he was not finally unaware of the fact that a body might fly off from the circular course at a tangent, in general he was perhaps a little too "Copernican" even in his mechanics—a little too ready to regard circular motion as the "natural" kind of motion, the thing which did not require to be explained. In reality, under the terms of the new physics, it was precisely this circular motion which became "violent" motion in the Aristotelian sense of the word. The stone that is swung round in a sling requires a constant force to draw it to the centre, and needs the exertion of violence to keep it in a circular path and prevent it from flying off at a tangent.

The men who succeeded Galileo made a cleaner affair of this business of geometrising a problem, and drew their diagrams in a space more free, more completely empty, and more thoroughly neutral. We can see at times how the new science had to dispose of mental obstructions in the achievement of this task, as when the two vertical sides of a balance were assumed to be parallel and the objection was raised that they must meet at the centre of the earth. It was easy to reply: "Very well, let us leave the centre of the earth out of the picture, let us suspend the balance up in the sky, far above the sun itself. Let us take it even an infinite distance away if necessary. Then we can be satisfied that the lines are really parallel." If there was a threat that the diagram should be spoiled by the operation of gravity they would say: "Away with gravity! Let us imagine the body placed in heaven, where there is neither up nor down—where up and down, in fact, are as indifferent as right and left." It was possible to argue: "Surely God can put a body in totally empty space, and we can watch it moving where there is nothing in the universe to attract or repel or in any way interfere with it."

The Aristotelian system had never been conducive to such a policy, which was necessary for the "geometrising" of problems, and which rendered science itself more amenable to a mathematical mode of treatment. It had not even been conducive to such a simple thing as "the parallelogram of forces," though Simon Stevin may not have been absolutely original when he produced this device while Galileo was still a young man. The Aristotelian system had discouraged the idea of the composition of motions, and was uncongenial to any mathematical treatment of the path which a body would follow when one motion happened to be complicated by another. We have seen how, in the case of projectiles, the Peripatetics had been unwilling to consider a mixture of motions, and had preferred to regard the body as driving forwards in a straight line until that motion was spent, and then quickly turning to drop vertically to the ground. It had been the new school which had begun to curve the path of the moving body and produced the view that in the mathematical world (which for a time they confused with the real world) the projectile described a parabola. And they worked out by mathematics the angle at which a gun must stand in order to fire the farthest; leaving their conclusion to be tested afterwards by actual experiment. All this helps to explain why Galileo could be in the position of defending what he called the mathematical method even against the experimental system of the better Aristotelians. It helps to explain also why Sir Francis Bacon, for all his love of experiments, was in a certain sense inadequate for the age, and proved to be open to criticism in the seventeenth century because of his deficiency in mathematics. In a certain sense he saw the importance of mathematics —the necessity of making calculations on the results of experiments in physics, for example—and on one occasion he even made an emphatic statement in regard to this matter. What he lacked was the geometer's eye, the power to single out those things which could be measured, and to turn a given scientific problem into a question of mathematics.

It was the extension of the new method that was to prove exceptionally important, however. Having conceived of motion in its simplest form—motion as taking place in this empty directionless space where nothing whatever could interfere with it and no resisting medium could put a check on it—the modern school could then reverse the process and collect back the things they had thrown away. Or, rather, we must say, they could draw more and more of these things into their geometrised world and make them amenable to the same kind of mathematical treatment. Things like air-resistence, which had been read out of the diagram at the first stage of the argument, could now be brought back into the picture, but brought back in a different way—no longer as despots but as subjugated servants. These things themselves were now caught into the mathematical method and turned into problems of geometry; and the same mode of treatment could be applied to the problem of gravity itself. The very method which the new science had adopted was one that directed the mind to more fields of enquiry and suggested new lines of experiment—attracting the student to things that would never have caught the attention of the Aristotelian enquirer. And the new avenues which were opened up in this way, even for experiment, were to carry the natural sciences away from that world of common-sense phenomena and ordinary appearances in which not only the Aristotelians but also the theorists of the impetus had done so much of their thinking. In particular, the mind was to be constantly directed in future to those things—and was to apply itself to those problems—which were amenable to measurement and calculation. Galileo therefore spoke very much to the point when he said that shape, size, quantity and motion were the primary qualities which the scientist should seek to examine when he was enquiring into given bodies. Tastes, colours, sound and smells were a matter of comparative indifference to him—they would not exist, he asserted, if human beings had not possessed noses and ears, tongues and eyes. In other words, science was to confine its attention to those things which were capable of measurement and calculation. Other objects which might be unamenable to such mathematical treatment in the first instance might still in the course of time be resolved into the same fundamentals. They might be translated or transposed into something else, and so, at a later stage of the argument, might become capable of being measured and weighed in turn.

In any case, it is essential that our interest in the experiment method as such should not cause us to overlook a matter of which the seventeenth century itself was clearly conscious—namely, the importance of mathematics in the developments that were taking place. When the interpretation of the whole scientific revolution is in question, certain facts which seem to have a bearing upon this issue strike the outsider as peculiarly significant. We have already met with a number of important aspirations and developments that belong to the fifteenth and sixteenth centuries—hints of a more modern kind of mechanics for example, foreshadowings of analytical geometry, discussions which seem to point towards what we call mathematical physics, and even intuitions concerning the value of the purely quantitative method in the natural sciences. We are told, however, that these interesting developments were brought to a halt, apparently because the middle ages lacked the necessary mathematics—the world had to wait until more of the mathematics of the ancient world had been recovered at the Renaissance. It would appear that there can exist a case of what might be called stunted development in the history of science. A movement may be checked, almost before it has cut any ice, if one of the requisite conditions happens to be lacking for the time being. In a similar way, we learn that Kepler's discovery of the laws of planetary motion was made possible only by the fact that he inherited and developed further for himself the study of conic sections, a study in which he was famous in his day. And certainly Tycho Brahe's astronomical observations became a revolutionary factor in history only when the mathematical mind of a Kepler had set to work upon that collection of materials. At a later date the same phenomenon recurs and we learn that the problem of gravitation would never have been solved—the whole Newtonian synthesis would never have been achieved—without, first, the analytical geometry of René Descartes and, secondly, the infinitesimal calculus of Newton and Leibnitz. Not only, then, did the science of mathematics make a remarkable development in the seventeenth century, but in dynamics and in physics the sciences give the impression that they were pressing upon the frontiers of the mathematics all the time. Without the achievements of the mathematicians the scientific revolution as we know it, would have been impossible.

It was true in general that where geometrical and mathematical methods could be easily and directly applied—as possibly in optics—there was very considerable development in the seventeenth century. In the period we have now reached—in the age of Galileo—arithmetic and algebra had attained something like their modern external appearance—the Frenchman, François Viète, for example, had established the use of letters to represent numbers; the Fleming, Simon Stevin, was introducing the decimal system for representing fractions; various symbols, now familiar to students, were coming into use between the fifteenth century and the time of Descartes. At the same time aids to mathematical calculation—a matter of importance to students of the heavenly bodies—were being created, such as John Napier's logarithms, developed between 1595 and 1614, and other devices for simplifying multiplication and division—the "bones," for example, which in the seventeenth century would appear to have had greater renown even than his logarithms. It has been pointed out that as algebra and geometry had developed separately—the former amongst the Hindus and the latter amongst the Greeks—the marriage of the two, "the application of algebraic methods to the geometric field," was "the greatest single step ever made in the progress of the exact sciences." The crucial development here came to its climax in the time of Descartes. Descartes put forward the view that sciences involving order and measure—whether the measure affected numbers, forms, shapes, sounds or other objects—are related to mathematics. "There ought therefore to be a general science—namely, mathematics," he said, "which should explain all that can be known about order and measure, considered independently of any application to a particular subject." Such a science, he asserted, would surpass in utility and importance all the other sciences, which in reality depended upon it. Kepler said that just as the ears are made for sound and the eyes for colour, the mind of man is meant to consider quantity and it wanders in darkness when it leaves the realm of quantitative thought. Galileo said that the book of the universe was written in mathematical language, and its alphabet consisted of triangles, circles and geometrical figures. There is no doubt that, in both Kepler and Galileo, Platonic and Pythagorean influences played an important part in the story.

If all these things are kept in mind we can see why the

resort to experiment in the natural sciences now came to have direction, came at last to be organised to some purpose. For centuries it had been an affair of wind and almost pointless fluttering—a thing in many respects irrelevant to the true progress of understanding—sometimes the most capricious and fantastic part of the scientific programme. There had been men in the middle ages who had said that experiment was the thing that mattered, or had realised that behind the natural philosophy of the Greeks there had been experiment and observation in the first place. But that was not enough, and even in the seventeenth century a man like Sir Francis Bacon, who harped on the need for experiments but failed to hitch this policy on to that general mathematising mode of procedure which I have described, was early recognised to have missed the point. In the thirteenth century, a writer called Peregrine produced a work on the magnet, and many of his experiments prepared the way for the remarkable book on the magnet produced by William Gilbert in 1600. The chief influence that came from Gilbert's book, however, emerged from his cosmic speculations based on the thesis that the earth was itself a great magnet, and Sir Francis Bacon was ready to seize upon the fact that this was not a hypothesis demonstrated by experiment, the thesis did not arise in the appointed way out of the experiments themselves. Even Leonardo da Vinci had tended to cast around here and there, like a schoolboy interested in everything, and when he drew up a plan of experiments in advance—as in the case of his projected scheme of study on the subject of flying—we can hardly fail to realise that here are experiments, but not the modern experimental method. Neither the medieval period nor the Renaissance was lacking in the ingenuity or the mechanical skill for modern technical achievement, as can be seen from the amazing contrivances they produced even where no urgent utilitarian purpose provided the incentive. Yet it is not until the seventeenth century that the resort to experiments comes to be tamed and harnessed so to speak, and is brought under direction, like a great machine getting into gear.

Even when one is interested in the scientific revolution

primarily as a transformation in thought, one cannot ignore those wider changes in the world which affect man's thinking or alter the conditions under which this thinking takes place. It is coming to be realised that the history of technology plays a larger part in the development of the scientific movement than it was once understood to do; and in fact the history of science is bound to be imperfect if it is regarded too exclusively as the history of scientific books. Some of the influence of industry and engineering upon scientific thought is difficult to locate, as yet, and might well be difficult to prove. But apart from the transference of ideas and techniques, there must have been an appreciable effect of a subtle kind upon the way in which problems were tackled and upon man's feeling for things, his feeling perhaps even for matter itself. A series of famous sixteenth-century books has put on record the technical progress which had then been achieved in various fields—in mining and metallurgy, for example; and some of this work must be regarded as preparing the way for modern chemistry, which it would be wrong to imagine as springing out of alchemy alone. On this technical side, and especially in the field of mechanics and hydrostatics, there is no doubt that Archimedes had a further influence on the course of the scientific revolution—we may almost regard him as the patron saint of the mechanically-minded and of the modern experimenters in physics. At first there was a considerable gulf between the practical man and the theorisers. The navigators would be too ignorant of mathematics, while the mathematicians lacked any experience of the sea. Those who worked out the trajectory of projectiles, or the appropriate angle of fire, might be far removed from the men who actually fired the guns in time of war. The map-makers, the surveyors, the engineers had long required some mathematics, however; the Portuguese discoverers had needed science to help them when they sailed south of the equator; William Gilbert had associated with navigators; and Galileo speaks of the kind of problems which arose in the ship-building yards at Venice, or in connection with the handling of artillery, and the pumping of water in mines. Indeed, it is proper to picture Galileo as passing his time in a sort of workshop with trained mechanics as his assistants, for ever making things—even making things for sale—and carrying out experiments, so that in him the mechanic or artisan combined with the philosopher to produce a modern type of scientist.

It has been argued that the growing number of mechanical objects in the world at large had induced also a sort of specialised interest or a modern attitude of mind—an interest in the sheer question of the way in which things worked, and a disposition to look upon nature with the same preoccupation. Apart from the famous cases in which a strategic experiment might bring the solution of a particular problem, Galileo gives the impression of having experimented so constantly as to gain an intimacy with movement and structures—he has watched the ways of projectiles, the operation of levers and the behaviour of balls on inclined planes, until he seems to know them, so to speak, from the inside in the way that some men know their dogs. And clocks worked by wheels were still a surprisingly new thing in the world when there appeared in the fourteenth century the suggestion that the heavenly bodies might be like a piece of clockwork. The early propaganda on behalf of the scientific movement laid remarkable stress on the utilitarian results that were expected from it; and this was one of the grounds on which the scientists or the scientific societies called for the patronage of kings. Sometimes there seems to be a curious correspondence between the technical needs of the age and the preoccupations of scientific enquirers, even when the precise connection escapes us or is hard to locate—as in the case of ballistics in the sixteenth century and hydraulic problems, perhaps, in the seventeenth. Much of the attention of the Royal Society in its early years was actually directed to problems of practical utility. And for a remarkably long period one of the topics constantly presented to the technicians and scientists was a matter of urgent necessity—the question of the finding of a satisfactory way of measuring longitude. It is not surprising that much of the work of students in our time has been turned to the history of technology.

One thing becomes significant in the seventeenth century and that is the creation of scientific instruments, especially measuring instruments; and it is hard for us to realise how difficult things must have been in earlier centuries without them. The telescope and the microscope appear at the very beginning of the century—and may have been devised a little earlier—and it is difficult not to regard them as a by-product of the glass- and metal-polishing industries in Holland. The microscope proved to be inadequate, however, for a long time, owing apparently to a defect, not in industrial technique as such, but in the actual science of optics. A more powerful single lens was produced, however, in the middle of the century, and much of the important work in the later period was really done with that. Galileo represents an important stage in the development of the thermometer and the pendulum-clock; and the barometer appears in the middle of the century; but for a long time it was possible to detect just the fact that the temperature was changing without having a reliable scale for the actual measurement of temperature. A really accurate thermometer did not exist until the eighteenth century. In the middle of the seventeenth century, again we meet with the momentous discovery of the air-pump, and only after this time do we see the use of the blow-pipe in chemical analysis. Van Helmont in the earlier half of the century studied gases, invented the word gas, and found that different kinds of gases existed—not simply air—but he was greatly handicapped, as he had no means of collecting and isolating a particular gas that he might want to examine, nor did he achieve our modern conception of what is a "gas." When one considers the richness and the fantastic nature of the objects that littered the laboratory of the alchemist even in the sixteenth century, one may feel that it can hardly have been the lack of industrial technique which delayed the appearance of some of the modern scientific instruments; though it appears that where purity or accuracy was highly necessary, either in the glass or in the metal-work, the technical progress achieved by the seventeenth century is a factor that affects the case. We may gather from repeated statements in books and correspondence that the experimental method in the first half of the seventeenth century involved a serious financial burden on its practitioners. Later in the century, when the informal gatherings of scientific workers turned into scientific societies—the Royal Society in England, the *Académie des Sciences* in France (and similar bodies earlier still in Italy), these societies helped to bear the expense of experiments. Their publications, and the establishment of a periodical literature, speeded up still more the communication and collation of scientific results. It would seem not to have been until the middle of the century that scientific publications really took the form of the communication of actual experiments. Sometimes, as in the works of Galileo, a point would be demonstrated by reasoning, though possibly it had been discovered in the course of experiment first of all.

Chapter 6

Bacon and Descartes

IT IS COMPARATIVELY EASY for people today to accommodate their minds to changes that may take place in upper regions of the different sciences—changes which from year to year may add further weight to the curriculum of the undergraduate student of the subject. It is not clear what the patriarchs of our generation would do, however, if we were faced with such a tearing-up of the roots of science that we had to wipe out as antiquated and useless the primary things said about the universe at the elementary school—if we had even to invert our attitudes, and deal, for example, with the whole question of local motion by picking up the opposite end of the stick. The early seventeenth century was more conscious than we ourselves (in our capacity as historians) of the revolutionary character of the moment that had now been reached. While everything was in the melting-pot—the older order undermined but the new scientific system unachieved—the conflict was bitterly exasperated. Men were actually calling for a revolution—not merely for an explanation of existing anomalies but for a new science and a new method. Programmes of the revolutionary movement were put forward, and it is clear that some men were highly conscious of the predicament in which the world now found itself. They seemed to be curiously lacking in discernment in one way, however, for they tended to believe that the scientific revolution could be carried out entirely in a single lifetime. It was a case of changing one lantern-slide of the universe for another, in their opinion—establishing a new system to take the place of Aristotle's. Gradually they found that it would need not merely one generation but perhaps two to complete the task. By the close of the seventeenth century they had come to see that they had opened the way to an indefinitely expanding future, and that the sciences were only in their cradle still.

Before the seventeenth century had opened, the general state of knowledge in regard to the physical universe had been conducive to the production of a number of speculative systems—these not founded upon scientific enquiry as a rule, but generally compounded out of ingredients taken from classical antiquity. Already in the sixteenth century, also, attention had been directed to the question of a general scientific method, and in the seventeenth century this problem of method came to be one of the grand preoccupations, not merely of the practising scientist, but, at a higher level, amongst the general thinkers and philosophers. The principal leaders in this seventeenth-century movement were Francis Bacon in the first quarter of the century, who glorified the inductive method and sought to reduce it to a set of regulations; and Descartes, whose work belongs chiefly to the second quarter of the century and who differed from Bacon not only in his glorification of mathematics as the queen of the sciences, but in the emphasis which he placed on a deductive and philosophical mode of reasoning, which he claimed to have screwed up to such a degree of tightness that it possessed all the discipline and certainty of mathematical reasoning. In the time of Newton and well into the eighteenth century, there was a grand controversy between an English school, which was popularly identified with the empirical method, and a French school, which glorified Descartes and came to be associated rather with the deductive method. In the middle of the eighteenth century, however, the French, with a charm that we must describe as Mediterranean, not only submitted to the English view of the matter, but in their famous *Encyclopédie* made even too ample a return, placing Bacon on a pedestal higher perhaps than any that had been given him before. It would appear that their excess of graciousness or charity brought some confusion into historical science at a later stage in the story.

Attacks on Aristotle had been increasingly common and sometimes exceedingly bitter in the sixteenth century. In

1543—a year which we have already seen to be so important in connection with Copernicus and Vesalius as well as the revival of Archimedes—Pierre Ramus produced his famous *Animadversions on Aristotle.* This work, which was known to Francis Bacon, and which attacked Aristotle without ever really understanding him, proposed an alternative method which was rather that of a humanist and professor of Belles Lettres—namely, studying nature through the best writers, and then applying deductive and syllogistic procedures to the result. In 1581 another writer, François Sanchez, produced a further attack on Aristotle, and more particularly on the modern followers of Aristotle —a work which provides a remarkable anticipation of Descartes. He said:

> I questioned the learned men of bygone centuries; then I consulted those who were my contemporaries . . . but none of their replies was satisfactory . . . So I turned in upon myself & put everything to doubt, as though I had never been told anything by anybody. I began to examine things myself in order to discover the true way of gaining knowledge—Hence the thesis which is the starting-point of my reflections: the more I think, the more I doubt.

He attacked the syllogistic reasoning of the prevalent Aristotelian school, because it turned men away from the study of reality and encouraged them to play a sophistical game of verbal subtlety. He promised to expound the true method of science, but in the fifty years of life that were left to him he never fulfilled the promise. One participant in the controversies over scientific method, Everard Digby, was teaching Logic in the University of Cambridge when Francis Bacon was there in his youth; and a German scholar has shown that at certain points Bacon appears to have followed the ideas of this man.

Bacon held that if Adam, owing to the Fall, had lost for the human race that domination over the created world which it had originally been designed to possess, still there was a subordinate command over nature, available if men worked sufficiently hard to secure it, though this had been thrown away by human folly. There had been only three short periods of genuine scientific progress throughout the whole course of human history, he said—one in Greek times, one in the Roman period, and the third which was being enjoyed in the seventeenth century. In each of the two ancient periods the era of scientific progress had been confined to two hundred years. The earlier Greek philosophers had set the course of enquiry on the right lines, but Plato and Aristotle had supervened, and they had come to prevail precisely because, being of light weight, they had managed to ride much farther down upon the stream of time. They had survived the storms of the Barbarian Invasions precisely because they had been shallow and buoyant, and Aristotle, in particular, had owed his remarkable sway in the world to the fact that, like the Ottoman sultans, he had pursued the policy of destroying all rivals. As for the scholastics of the middle ages, they had had "subtle and strong capacities, abundance of leisure, and but small variety of reading, their minds being shut up in a few authors"; and therefore they had "with infinite agitation of wit, spun out of a small quantity of matter those laborious webs of learning which are extant in their books.' Bacon was impressed by the fact that scientific knowledge had made such extraordinarily little progress since the days of antiquity. He begins by saying that men ought to "throw aside all thought of philosophy, or at least to expect but little and poor fruit from it, until an approved and careful natural and Experimental History be prepared and constructed."

> For to what purpose are these brain-creations and idle display of power. . . . All these invented systems of the universe, each according to his own fancy [are] like so many arguments of plays . . . every one philosophises out of the cells of his own imagination, as out of Plato's cave.

He uses the term "history" in the sense that we have in mind when we speak of natural history, and he regards it as comprising a collection of data, the fruits of enquiry.

He believed that many men had been led away by allowing their scientific work to become entangled in a search for final causes, which really belonged rather to philosophy, and which he said corrupted the sciences, except those relating to the intercourse of man with man. In education he thought that scholars were introduced too early to logic and rhetoric, which were the cream of the sciences since they arranged and methodised the subject-matter of all the others. To apply the juvenile mind to these before it had been confronted with the subject-matter of the other sciences was like painting and measuring the wind, he said —on the one hand it degraded logic into childish sophistry, on the other hand it had the effect of making the more concrete sciences superficial. In his reaction against the older ways of discussing science, Bacon carried the attack beyond the bounds of prudence on occasion—denying the value of syllogistic modes of reasoning in a way that the modern philosopher would disapprove of; though the general line of attack was understandable, and very useful in view of the situation of things at that time. Bacon wanted men to close in on nature and get to grips with her, bringing their minds to mix in its actual operations. "The secrets of nature," he said, "betray themselves more readily when tormented by art than when left to their own course." "It is best to consider matter, its conformation, and the changes of that conformation, its own action, and the law of this action in motion." He did not support a dead kind of empiricism; the empirics, he said, were like ants merely heaping up a collection of data. The natural philosophers still generally current in the world, however, were rather like spiders spinning their webs out of their own interior. He thought that the scientists ought to take up an intermediate position, like that of the bees, which extracted matter from the flowers and then re-fashioned it by their own efforts. Existing interpretations of nature, he said, were generally "founded on too narrow a basis of experiment." "In any case," he insisted, "the present method of experiment is blind and stupid"—men did it as though they were schoolboys engaged "as it were in sport." He talked of "desultory, ill-combined experiment." The alchemists, he said, had theoretical preconceptions which hindered them from either carrying out their experiments along useful lines or extracting anything important from their results. Men in general glanced too hastily at the result of an experiment, and then imagined that the rest could be done by sheer contemplation; or they would fly off into the skies with a hasty first impression and attempt to make this square with the vulgar notions already existing in their minds. Even Gilbert working on the magnet had no unity or order in his experiments—the only unity in his treatise lay in the fact that he had been ready to try out anything that there was to try out with a magnet.

Now it was Bacon's firm principle that if men wanted to achieve anything new in the world, it was of no use attempting to reach it on any ancient method—they must realise that new practices and policies would be necessary. He stressed above all the need for the direction of experiments—an end to the mere haphazard experimenting—and he insisted that something far more subtle and far-reaching could be achieved by the proper organisation of experiments. It is quite clear that he realised how science could be brought to a higher power altogether by being transported away from that ordinary world of commonsense phenomena in which so much of the discussion had hitherto been carried on. He insisted on the importance of the actual recording of experiments, a point which, as we have already seen, was now coming to be of some significance. He urged that experimenters in different fields should get together, because they would knock sparks off one another; and things done in one field would give hints to people working in another field. In this sense he anticipated the point of Professor Whitehead who shows how, precisely in this period, the knowledge of several different branches of science at once might have an enriching effect on each. Also, suggestions which are scattered in various parts of Bacon's work seem to have served as an inspiration to some of the men who founded the Royal Society.

It often happens that when the philosopher comes to deal with the position of a man like Bacon in the history of thought, he lays great stress either upon the internal inconsistencies that may exist in the intellectual system in question, or on the actual correctness—from a modern point of view—of the man's conclusions, which in the present case would mean the correctness of Bacon's predictions concerning the character and the method which modern science was going to take upon itself. A modern critic may lay about him right and left on the subject of the philosophy of the nineteenth-century Utilitarians, if that teaching merits the name of philosophy; but the historian who remembers all the inhibitions that restricted parliamentary action at the beginning of the nineteenth century, and who has in mind the vast flood of legislation that began to appear in the second quarter of that century, can hardly help realising that on a lower level altogether—in a sub-philosophical field—it required a first-class campaign to get rid of the inhibitions and to persuade people of the commonplace fact that laws could be regarded as mere ministers to ordinary utility, that anachronistic legislation was not a thing to be preserved for semi-mystical reasons. It is at this lower level of analysis—in this sub-philosophical realm—that Bacon is so interesting and so important in history, and we must not ask ourselves: How many people adopted the Baconian system literally and *in toto?* We must not be surprised that even in the seventeenth century it was precisely the people in the same line of thought as Bacon—the logicians—who were the least influenced by his teaching. We must not be disconcerted if even at the very heart of his teaching, where he purported to show exactly how the results of experiments could be turned into generalisations, he was on occasion less original than he intended to be, and on occasion actually mistaken. In the days when the grand campaign against Aristotle was coming to its height he produced a programme and manifesto, and some of the most important things that he said are dead to us but were quivering with life in the seventeenth century, because they were right and so happen to have become commonplaces today. He did not produce Baconians taking over his whole system, but rather stimulated people in a piecemeal way—people who apparently did not always even read his works in their entirety. And since authors who merely write about method are liable to mistakes which are avoided by men who are actually engaged in research (for the simple reason that the latter can often hardly help following their noses half the time), it is not surprising if some people thought they were disciples of his method when in reality they were doing something different, something which in many cases would be better still. In his own words, "he rang the bell which called the wits together," and many of his aphorisms —especially where he is diagnosing the causes of common errors in thought—would give both profit and stimulus to students of history today. Paradoxically enough, there is possibly some truth in the view that the Baconian influence has been most direct in some of what might be called the literary sciences.

He has been attacked because there is so much in his writing that savours of the old Aristotle; but that was necessary since his system ranged over all the realms of thought and philosophy. He has been mocked because so many of his beliefs about nature were still medieval—but that was also true of the various scientists of the time. If he believed in the existence of vital spirits in the blood, so did William Harvey himself, as we have seen. If he described inanimate things as having aspirations and dispositions, or as being drawn by affection to one another, Robert Boyle, much later in the century, explicitly defended this mode of expression. He has been criticised because when he collected data he included fables and old wives' tales along with established scientific facts. He instructed scientific workers to examine the fables, however, and repeatedly he made the point that he expected to find his data corrected by enquiries that would take place in the future. When he set out to provide a starting-point for scientific enquiry, and to assemble his catalogues of known facts, achieved experiments and suggested hypotheses, he made terrible mistakes, for he was writing before modern physics or chemistry or astronomy or physiology had really begun to be established. The mistaken science of the past always appears as blind superstition to the future, and Bacon at one point and another would fail to free himself from existing prejudices or, alternatively, to prevent his mind from running to fantastic conjecture. But he realised the possibility of error in advance, and said that it mattered little if his experiments were wrong, "since it must needs happen in beginnings." He claimed that at any rate his compendiums were more useful than the scientific knowledge that had hitherto been available. He constantly reiterated, furthermore, that he intended only to offer hypotheses for people to examine; even if they were wrong they would be useful, he said. On one occasion he noted that it was too early to put forward an opinion on a given issue, but he would offer his own for the time being because it might seem like cowardice if he did not. On another occasion he said:

> I do not pronounce upon anything, I set down and prescribe but only provisionally . . . I sometimes make attempts at interpretation . . . [but] what need have I of pride or imposture seeing that I so often declare that we are not furnished with so much history or experiments as we want and that without these the interpretation of nature cannot be accomplished; and that therefore it is enough for me if I set the thing on foot.

If we look for the root of the error that was in him—the cause that was perhaps behind the other causes—it lay in his assumption that the number of phenomena, the number even of possible experiments, was limited, so that the scientific revolution could be expected to take place in a decade or two. "The particular phenomena of the arts and sciences are in reality but as a handful," he once said; "the invention of all causes and sciences would be the labour of but a few years." He thought that he could make catalogues of facts, or required experiments and of suggested hypotheses; and while on the one hand he imagined that the whole renovation of the sciences would be held up unless he provided this guide-book, he spoke at times as though, once his compendium had been compiled, the work of science would proceed almost by rule of thumb.

Even here he was not so inelastic as some people have made out, however, and not so blind to the importance of hypotheses. If he thought it his special function to provide the hypotheses, he would add the remark that further ones would suggest themselves to the enquirer as he went along.

He believed that out of experiments one could draw generalisations, and that these generalisations themselves would point the way to further experiments. In a curious but significant way he seems to have foreseen the structure that science was to take in the future—a point which may best be illustrated perhaps by adapting an example suggested in a lecture by Professor Broad. Bacon thought that at the first immediate level the generalisations or axioms which might be drawn out of experiments were too low-grade, too near to concrete facts to be of any great utility. Knowledge is limited if we only know that heat can be produced by mixing sulphuric acid and water; and the knowledge is of little value unless these two substances happen to be at hand. The very highest generalisations of all, however, are out of reach, too near to God and to final causes; they must be left to the philosopher. The intermediate axioms are the ones that are "true, solid and full of life," says Bacon—the rather higher generalisations which can be reached by the method of climbing up to them from below. If one knows that violent molecular motion is the factor that produces heat, one is in possession of a wider form of generalisation and this will greatly increase one's power over nature. Incidentally, Bacon makes the remark that there are some things which have become so familiar or which are accepted so automatically that people take them as self-evident, though they are just the things which are most in need or re-examination. In this connection he specifies the causes of gravity, the rotation of the heavenly bodies, heat, light, density and organic formation. He shows some insight in recognising that the progress of science would consist in the pursuit of enquiries upon lines such as these.

It was on the mathematical side—and particularly, so to speak, on the geometrical side—that Bacon missed the point of that kind of science which was to spring from Galileo. His error ought not to be exaggerated. He says in one place: "The investigation of nature is best conducted when mathematics are applied to physics." He says in other place: "If physics be daily improving, and drawing out new axioms, it will continually be wanting fresh assistance from mathematics." On the other hand, he regarded mathematics merely as the hand-maid to physics, and actually complained of the dominion which it was beginning to exercise in that science. It was all very well to do sums on the results of one's experiments, but Bacon specifically disliked Galileo's method of turning the problem of motion, in the way we have seen, into the problem of geometrical bodies moving in geometrical space. Far from wanting to read away the air-resistance, in the way the new school of scientists were doing, he wanted to add other things to the picture—for example, the tensions that were bound to take place within the moving body itself. Far from wanting to abstract and to isolate any aspect of a scientific problem, so that motion could be considered as a line drawn in geometrical space, he longed rather to load all the concreteness back into the problem, to see a picture which included air-resistance and gravity and the internal texture of the body itself. Even in the case of the celestial bodies he deprecated the purely geometrical study of motion and said that the enquirer ought not to overlook the question of the kind of material out of which the planets were manufactured. On the subject of projectiles he declined to accept either Aristotle's theory that the motion was caused by the rush of air, or the impetus-theory which had hitherto been its principal rival. He put forward the hypothesis that if motion continued after an impact this was the result of the play of the internal forces and stresses which had been put into operation by the shock of the original percussion.

Indeed, it is important in the study of Bacon not merely to know the skeleton of his system, but to observe how he treats the problems in any of the branches of science. And it is not sufficient to note whether he was right or wrong according to the views of the present day. We must know where each particular science stood at the time when he was writing, and exactly how he would play upon the margin of it. There is one field in which this matter may perhaps be usefully discussed at the present moment, since it is connected with problems which we have already traversed in a general way; and that is, the field that relates to the problem of the skies. It is the more interesting from the fact that Bacon is so often summarily dismissed for his anti-Copernican prejudices.

On this subject Bacon begins by saying:

> I will myself therefore construct a Theory of the Universe according to the measure of the history, [the established facts,] as yet known to us; keeping my judgment however in all points free, for the time when history, and by means of history, my inductive philosophy shall have been further advanced.

Later he says:

> Nevertheless I repeat once more that I do not mean to bind myself to these; for in them as in other things I am certain of my way but not certain of my position. I have introduced them by way of interlude lest it be thought that it is from vacillation of judgment or inability to affirm that I prefer negative questions.

He says that many astronomical systems can be put forward which will cover the phenomena. The Ptolemaic is one, the Copernican is another. Either will account for the observed movements, but Bacon prefers the system of Tycho Brahe, the intermediate system by which some of the planets go round the sun and these all together go round the motionless earth. He regrets, however, that Tycho Brahe had not worked out the mathematics of such a system and shown its operation in detail. "Now it is easy to see," he says, "that both they who think the earth revolves and they who hold the primum mobile and the old construction are about equally and indifferently supported by the phenomena." He prefers, however, the view that the earth is stationary—"for that I now think the truer opinion," he says. Still, he puts the question as one for the reader to answer: Whether there is a system of the universe with a centre, or whether the particular globes of earth and stars are just scattered and dispersed, each, as he says, "on its own roots," or each as "so many islands in an immense sea." Even if the earth revolves it does not necessarily follow that there is no system of the universe, he says; for these are planets that do revolve round the sun. But though the rotation of the earth is an ancient idea, the Copernican view that the sun stands immovable at the centre of the universe is one which Bacon considers to be unprecedented. He is prepared to ask whether there may not be many different centres of the universe, the heavenly bodies being congregated in bundles or groups, so that he can picture them as separate parties of people each doing a separate dance. He addresses himself to the problem we discussed in connection with the modern doctrine of inertia when he says: "Let no one hope to determine the question whether the earth or heaven revolve in the diurnal motion unless he have first comprehended the nature of spontaneous rotation." In one place he makes it clear that he dislikes the movement of the earth because it would leave nature without any quiet, any immobility. Repeatedly he tells us that so far as the mathematical aspect is concerned the Copernican system is satisfactory, but he stumbles at the obstacle which we have seen to be the general difficulty even in the days of Galileo: the Copernican hypothesis has not yet been made to square with what is known of physical science in general. Bacon repeats that the mathematician-astronomers can never solve the problem by themselves. Let the observation of the heavenly bodies proceed—we are all the better if we can get the geometry of the skies correct—and the mathematical side of the work must certainly be dovetailed into the discoveries of physical science. On the mathematical side things are going well at the moment, especially with the new optical instruments; but there must be greater constancy of observation, greater severity of judgment, more witnesses to confirm observations, and each particular fact must be tested in different ways. The real weakness still lies in the physics, however. The enquirer ought to have regard to the actual material the stars are made of, learn about the appetites and behaviour of the stuff itself, which must be fundamentally the same in all regions of the heavens. Bacon declines to accept the view that the heavenly bodies are formed of an immaculate substance free from change and exempt from the ordinary forces of nature. It was heathen arrogance, not the Holy Scripture, he says, which endowed the skies with the prerogative of being incorruptible. Also he tells us: "I shall not stand upon that piece of mathematical elegance, the reduction of motions to perfect circles." Dispersed through his work are many references to Galileo's telescopic discoveries. He accepts all the empirical data that these observations provide; but he does not accept Galileo's theories, though he does quote Galileo with approval for the view that the effect of gravity diminishes as one goes farther away from the earth. When he discusses the question of the tides, he says that on the supposition that the movement of the earth causes the tides, certain things will follow—not that he personally holds with Galileo's theory on this subject. His own view is that the farthest skies and stars move rapidly in a perfect circle, but that as we come down nearer to earth the heavenly bodies themselves become more earthy and they move in a more resistant medium. Things becoming more heavy and gross as we approach the mundane region, their motion slows down in proportion as they are nearer to earth and hold a lower place in the skies. What appears to be the motion of the planets in one direction is merely the optical illusion produced by the fact that they are so much behind the highest skies and the farthest stars; it merely represents a lag in that single circular movement which they are all supposed to share. Not only is the pace reduced, but the circular motion is departed from, as one comes lower down in the sky and nearer to the gross and material earth. The total result is to produce in the sky the effect of spirals, and Bacon affects to wonder why the spiral has never been thought of before, since it represents an initial circular motion constantly going off the circle as it descends to more turgid realms. In his view the tides are the last weak effects of the total revolution of the skies around the motionless earth.

That was Bacon's system of the universe, though as we have already seen, it was a mere tentative hypothesis and he did not consider that the time had yet come for the production of a general synthesis. It is clear, however, that from the point of view of that time his work was essentially stimulating, especially in the signs it gave of an extraordinary elasticity of mind; and that many people were influenced by it, through their work might not itself have a Baconian look at the finish—his influence tended to make men better than himself, make them something better than mere Baconians. The numerous translations of his works into French in the first half of the seventeenth century show that he aroused great interest across the Channel.

With René Descartes, who lived from 1596 to 1650, we meet a system of thought much more intensive and concentrated, and much more intricately interlocked. We shall find this man, like Galileo, reappearing in various aspects in the story of the scientific revolution, sprawling over the whole area that is left of the seventeenth century. What requires notice at the moment is merely the short treatise —a thing almost of pamphlet size—entitled *A Discourse on Method,* which is one of the really important books in our intellectual history. To the historian its greatest significance lies, not in its one or two philosophical passages or in the disquisition on mathematics, but in its aspect as just a piece of autobiography. In this aspect it influenced, not merely those who were to become Cartesian in philosophy, but the world in general.

It was written in the vernacular, and Descartes meant to address himself to the natural reason of men whose minds had not been perverted by the traditions of the schools. Those who read the *Discourse on Method,* not profoundly as philosophers but superficially in the way in which people do read books, will understand, better than the philosophers ever do, the importance and the influence of Descartes in general history. More important perhaps than anything the author intended is the manner in which the book was misunderstood; and Descartes himself complains not only in his letters but in this very book of the way in which he was being misunderstood already. He says in the *Discourse* that when he hears his own views repeated he finds them so changed that he cannot recognise or acknowledge them as his—a remark which must go straight to the heart of every author. He cries out against those people who think that they can master in a day the things which he had taken twelve years to think out. He explains in the *Discourse* how he had come to feel that all the sciences which he had been taught in his youth had really told him nothing—how the various opinions to which men in different parts of the world were attached were so often merely the result of custom and tradition. The book is vivid as a chapter of autobiography, written by a man who after much travail decided that he must sweep away all ancient opinions and start all his thinking over again.

Bacon had talked of the need of "minds washed clean of opinions," but Descartes went further in his determination to unload himself of all the teaching which had been transmitted from the ancient world, his determination to doubt everything and start naked once again, without any foothold whatever save the consciousness that I who do the doubting must exist—even though I may doubt whether I am doubting. Those who never understood the positive teaching of Descartes, and who could never have risen to his philosophy, appreciated this dramatic rejection of inherited systems and ideas. And though he himself said that the attempt to overthrow all tradition in this way was not a thing to be carried out by any and every man; though he cautioned against any imitation of the sceptics—for, in fact, he was only doubting in order to find a firmer basis for belief or certainty—still the influence of the policy of methodical doubt was in the long run to be most significant on the destructive side and in the realm of general ideas. The misunderstanding of Descartes was made more easy, because, in fact, he did not intend his *Discourse on Method* to be anything more than a mere preface to his real study and survey of the problem of method. The essay was an introduction to three treatises—the *Dioptric,* the *Meteors* and the *Geometry,* and it was the intention of Descartes to develop the idea of his method by illustrating it in action, showing how it operated in concrete cases—that is to say, in different branches of science. It proved to be these three treatises that provided the greater sensation and drew the chief attention at the time; but the world soon gets tired of reading out-of-date science, so that these parts of the work gradually lost their initial importance. The *Discourse on Method,* which is stimulating to read at any time, gradually detached itself from the essays to which it was a mere preface, and came to stand on its own feet.

Descartes believed that the essential capacity to see reason was distributed throughout the human race without any difference of degree, however clouded it might be by prejudice or by the illusions of the imagination. He established what became the great principle of common sense in modern times, for if he insisted on one point more than any other it was in his thesis: "All things which we clearly and distinctly conceive are true." If I say "I think, therefore I am," I am not really deducing anything—I am announcing a kind of intuitive perception of myself, a perception which nothing can get behind. Beyond that, if I say "I have a body," I am liable to be misled by pictures and fogs—the visual imagination is precisely the thing that is unreliable. The people who say "I believe in my body because I can see it clearly, but I cannot see God" were turning a popularised Descartes to the purpose exactly the reverse of what had been intended. In the system of Descartes God was another of those clear ideas that are clearer and more precise in the mind than anything seen by the actual eye. Furthermore, everything hung on this existence of a perfect and righteous God. Without Him a man could not trust in anything, could not believe in a geometrical proposition, for He was the guarantee that everything was not an illusion, the senses not a complete hoax, and life not a mere nightmare.

Starting from this point, Descartes was prepared to deduce the whole universe from God, with each step of the argument as clear and certain as a demonstration in geometry. He was determined to have a science as closely knit, as regularly ordered, as any piece of mathematics—one which, so far as the material universe is concerned (and excluding the soul and the spiritual side of things), would lay out a perfect piece of mechanism. His vision of a single universal science so unified, so ordered, so interlocked, was perhaps one of his most remarkable contributions to the scientific revolution. Indeed, he carried the unification so far that he said that one single mind ought to work out the whole system—he indulged at one time in the hope that he might carry out the whole scientific revolution himself. When others offered to help him with experiments he was tempted to reply that it would be much better if they would give him money to carry out his own.

The physics of Descartes, therefore, depends in a particular way upon his metaphysics; it provides merely the lower stages in an hierarchical system that definitely reaches back to God. Descartes is prepared to work out a whole system of the universe, starting with matter (or with what the philosophers call extension) on the one hand, and movement, purely local motion, on the other. Everything was to be accounted for mathematically, either by configuration or by number. His universe, granting extension and movement in the first place, was so based on law that no matter how many different universes God had created—no matter how different from one another these might be at the start—they were bound, he said, to become like the universe we live in, through the sheer operation of law upon the primary material. Even if God had created a different universe at the beginning, it would have worked itself round to the system that now exists. Even if He had made the earth a cube, it would have rolled itself into a sphere. Perhaps the most essential law in the physical system of Descartes was the law concerning the invariability in the amount of motion in the universe. Motion depended ultimately on God, and the law concerning the invariability in the amount of motion was a law which followed from the immutability of God. It might be thought that Descartes could have arrived at some such law by observation and experiment, or at least by taking it as a possible hypothesis and discovering that it actually succeeded, actually worked in practice. That would never have been sufficient for him, for it could never have provided that clinching demonstration, that exclusion of alternative possibilities, which it was the purpose of his system to achieve. What he wanted was the certainty of a deductive and quasi-geometrical proof, and he had to carry the question back to God, so that his physics had to depend on his metaphysics. Envisaging the matter with the eye of the geometer, however, and conceiving motion therefore so largely in its kinematic aspect, he laid himself open to the criticism that his system suffered from anæmia in respect of questions relating to dynamics. His law on the subject of the conservation of momentum proved unsatisfactory and had to be replaced by the law of the conservation of energy.

He tells us in the *Discourse on Method* that from one or two primary truths that he had established he was able to reason his way by the deductive method to the existence of the heavens, the stars and the earth, as well as water, air, fire, minerals, etc. When it came farther than that—to the more detailed operations of nature—he needed experiment to show him in which of the alternative ways that were possible under his system God actually did produce certain effects; or to discover which of the effects—amongst a host of possible alternatives that his philosophy would have allowed or explained—God had actually chosen to produce.

Experiment, therefore, only had a subordinate place in the system of Descartes, and in the later part of the seventeenth century the famous scientist Huygens, who criticised Bacon for his lack of mathematics, complained that the theories of Descartes were not sufficiently confirmed by experiment. The beauty and the unity of the system of Descartes lay in the fact that on the one hand it started from God and worked downwards by a system of reasoning that was claimed to be water-tight; while at the same time it worked upwards from below, drawing generalisations or axioms from the experiments. There are signs however, that Descartes would use an experiment to confirm a hunch or an hypothesis, but would close down the enquiry very soon—refusing to pursue further observations even when these might have affected the case in a more or less indirect manner. He worried much less about establishing a fact than about its explanation—his point was to show that, supposing this thing was a fact, his system would provide the explanation; and, indeed, this system would have explained the case supposing God at one point or another had taken an alternative course that might have been open to Him. So in his treatise on *Meteors,* which was one of the works attached to his *Discourse on Method,* he was prepared to explain how the clouds could rain blood, as was sometimes alleged, and how lightning could be turned into a stone. In fact, he confessed that he preferred to apply his method to the explanation of what were the ordinarily accepted phenomena, rather than to use experiment in order to find new phenomena or out-of-the-way occurrences. Many of his accepted "facts," like the ones I have just mentioned, were in reality taken over without examination from scholastic writers. He accepted the idea of the circulation of the blood, but quarrelled with Harvey concerning its cause and concerning the action of the heart. He said that when the blood was drawn into the heart it became so heated that it effervesced, caused the heart to expand, and leaped of its own motion into the arteries. In this case the truth was that he accepted unconsciously and without real examination the scholastic assumption that the heart functioned as the centre of heat.

The men who were influenced by Bacon were chiefly affected by the thesis that experiment was the thing that mattered in the natural sciences. And Robert Boyle, who shows clear marks of that influence, was criticised by Huygens and others for having built so little on the great number of experiments that he recorded. The founders of the Royal Society were under that general influence, and in the early proceedings of the Royal Society there is a rage for experiments, not only of what we should call the scientific kind, but in regard to curiosities and prodigies in nature, or in respect of invention and technological devices—sometimes experiments just to test old wives' tales.

In the synthesis of Descartes, however, as we shall see later, there is the economy and austerity of a highly concentrated deductive system. By its mechanisation it anticipated the structure that physical science was to assume in the future. But the combination of the mathematical and the experimental method in England was destined to put the natural science of Descartes into the shade before the seventeenth century had expired.

Chapter 7

The Effect of the Scientific Revolution on the Non-Mechanical Sciences

IT HAS ALREADY BEEN noted that the central studies in the scientific revelution were those of astronomy and mechanics. They represent the fields in which the most drastic changes and the most remarkable progress occurred in the seventeenth century. Picking up one end of the stick, we may consider the suggestion that astronomy might well have been ripe for such development because in so ancient a science observations had been accumulating for thousands of years and the process of revision was bound sooner or later to demand a new effort of synthesis. We may feel at the same time that the science of mechanics had an advantage in that it was a branch of study in which there was much to be achieved by simple devices, such as watching balls rolling down inclined planes. Picking up the other end of the stick, however, we may say that these sciences were spurred forward because in both cases there was an unusually bad hurdle to get over precisely at the point which the story had then reached. In the one case there was the difficulty of arriving at a proper conception of simple motion. In the other case there was the particular difficulty of conceiving or explaining the motion of the earth itself. With both the sciences it was to transpire that, once the hurdle had been surmounted, the way was left open to an astonishing flood of further change. Perhaps the mere development of mathematics and of the mathematising habit had much to do not only with the scientific revolution in general, but more especially with the surmounting of the significant hurdles that are here in question. It would seem to have been true in any case that the whole history of thought was to be affected by the new study of motion—whether on the earth or in the sky—which was the high-water mark of seventeenth-century science. That century was to see one attempt after another to explain many things besides motion, and, indeed, to interpret all the changes of the physical universe, in terms of a purely mechanistic system. The ideal of a clockwork universe was the great contribution of seventeenth-century science to the eighteenth-century age of reason.

It is never easy—if it is posible at all—to feel that one has reached the bottom of a matter, or touched the last limit of explanation, when dealing with an historical transition. It would appear that the most fundamental changes in outlook, the most remarkable turns in the current of intellectual fashion, may be referable in the last resort to an alteration in men's feeling for things, an alteration at once so subtle and so generally pervasive that it cannot be attributed to any particular writers or any influence of academic thought as such. When at the beginning of the sixteenth century an Englishman could write concerning the clergy that it was scandalous to see half the king's subjects evading their proper allegiance to the crown—escaping the law of the land—we know that he was registering a change in the feeling men had for the territorial state, a change more significant in that people were unconscious of the fact that anything novel had been taking place. Subtle changes like this—the result not of any book but of the new texture of human experience in a new age—are apparent behind the story of the scientific revolution, a revolution which some have tried to explain by a change in men's feeling for matter itself.

It is fairly clear in the sixteenth century, and it is certain in the seventeenth, that through changes in the habitual use of words, certain things in the natural philosophy of Aristotle had now acquired a coarsened meaning or were actually misunderstood. It may not be easy to say why such a thing should have happened, but men unconsciously betray the fact that a certain Aristotelian thesis simply has no meaning for them any longer—they just cannot think of the stars and heavenly bodies as things without weight even when the book tells them to do so. Francis Bacon seems unable to say anything except that it is obvious that these heavenly bodies have weight, like any other kind of matter which we meet in our experience. Bacon says, furthermore, that he is unable to imagine the planets as nailed to crystalline spheres; and the whole idea only seems more absurd to him if the spheres in question are supposed to be made of that liquid, ethereal kind of substance which Aristotle had had in mind. Between the idea of a stone aspiring to reach its natural place at the centre of the universe—and rushing more fervently as it came nearer home—and the idea of a stone accelerating its descent under the constant force of gravity, there is an intellectual transition which involves somewhere or other a change in men's feeling for matter. As we have already seen, there was a change also in the feeling that men had for motion, if only because Aristotle, thinking of simple motion, naturally had in his mind the picture of a horse drawing a cart, while the new age had good reason for focusing its primary attention on the projectile, which meant a difference in the apprehension of the whole affair.

In a similar way, a subtle intellectual change was giving people an interest in the operation of pure mechanism; and some have even said that this came from the growing familiarity with clocks and machines, though it would be impossible to put one's finger on any authentic proof of this. The great importance of astronomy and mechanics can certainly not be attributed to this alone, though such a factor may have helped to intensify the preoccupation of scientific enquiry with the question of mechanism. One thing is clear: not only was there in some of the intellectual leaders a great aspiration to demonstrate that the universe ran like a piece of clockwork, but this was itself initially a religious aspiration. It was felt that there would be something defective in Creation itself—something not quite worthy of God—unless the whole system of the universe could be shown to be interlocking, so that it carried the pattern of reasonableness and orderliness. Kepler, inaugurating the scientist's quest for a mechanistic universe in the seventeenth century, is significant here—his mysticism, his music of the spheres, his rational deity, demand a system which has the beauty of a piece of mathematics. There had been a time when men had sought to demonstrate God by means of miracles, and the intellect of man had yearned rather to discover the evidence of divine caprice in the world. At the moment which we have now reached, a difference of feeling marked a transformation in human experience; for it is clear that it was now rather the aspiration of the mind to demonstrate divine order and self-consistency. Without regularity in the ordinary workings of the universe the Christian miracles themselves could have no meaning. And, as we have seen, the aspiration to turn the created world into mechanism was part of the reaction against pan-psychic superstition, against the belief that nature itself was magical. It was clear that God could create something out of nothing, but it was obvious to Francis Bacon that nature would do nothing of the kind—the amount of matter in the universe must remain constant. We have already seen how Descartes deduced the conservation of momentum from his idea of the immutability of God.

It is perhaps not too much to say, therefore, that something in the whole intellectual climate of the age helps to explain the attempts which were made in this period to revive those systems which interpreted the nature of matter itself on purely mechanistic principles. It was this which led to the prevalence in the seventeenth century of various forms of what came to be called the corpuscular philosophy. The view became current that all the operations of nature, all the fabric of the created universe, could be reduced to the behaviour of minute particles of matter, and all the variety that presented itself to human experience could be resolved into the question of the size, the configuration, the motion, the position and the juxtaposition of these particles. The ancient atomic theories associated with Democritus and the Epicureans were brought to life again in a new context; but one broad difference existed—whereas the ancient theory had tended to attribute everything to the fortuitous combinations of atoms, so that the universe had been left, so to speak, at the mercy of chance, now there was assumed to be rationality in the mechanism itself—indeed, the corpuscular theories were the result of the search for rationality, and even part of the urge to justify God.

Francis Bacon, much as he protested against philosophical systems, took great pains to call attention to the importance of these atomic explanations of the universe. He has some interesting remarks upon this subject in a series of essays entitled *Thoughts on the Nature of Things.* He pictured the original atoms "throwing themselves into certain groups and knots"—their different combinations being sufficient explanation of the varieties of substance which are presented to the five senses of man. He saw the importance of the motion of the particles and suggested how many things—heat, for example—were to be explained by the mere fact of this motion, which took place on a minute scale inside the very fabric of solid substances. In this connection he pointed out that the great defect of the ancient thinkers was their failure to study and understand motion—a matter that was essential to the comprehension of the processes of nature. Some people believed that the minute particles which I have mentioned were the very last thing that could be reached in the analysis and subdivision of matter. They were hard and impenetrable and final—utterly incapable of further reduction. These people were prepared to desert a principle which had been accepted on the authority of Aristotle—prepared to admit the existence of a vacuum between the ultimate particles and inside the fabric of matter itself. They tended to follow Gassendi, who in 1626 announced his intention of restoring the philosophy of Epicurus, and produced a system specifically atomic in character. Others, who regarded a vacuum as impossible in any sense and believed therefore in the unbroken continuity of matter throughout the universe, tended rather to follow Descartes. On their view, matter was infinitely divisible, the particles could be broken up and, in fact, there was no really ultimate atom which represented the hard basis of all forms of substance. If all the air could be drawn out of a tube, that tube would still be as full as before, and the substance which it now contained would be continuous, however much more ethereal. There were men like Robert Boyle who were unwilling to decide between these two views, but still confessed their hankering after some form of what they called the corpuscular philosophy. In the seventeenth century the revelation of the intricate structure of nature—especially with the increasing use of the magnifying glass, the telescope and then the miscroscope—made people greatly interested in the minute subdivision of matter. Bacon himself is an illustration of the way in which men became aware of the extraordinary intricacy in the structure of things and the complexity of even the minute aspects of nature. The new philosophy enabled men to reduce the whole universe to matter and motion. It made possible the explanation of the whole of nature in mechanistic terms.

The intention—often explicitly formulated in the seventeenth century—of attempting to explain everything in the physical universe by mechanical processes had important effects upon the biological sciences, upon which it tended to imprint its own peculiar character. Such sciences would appear to have been stimulated by this new mode of treatment at the first stage of the story—stimulated to exceptional development on certain sides perhaps. It appears, however, that the time was to come when the mechanistic point of view became an embarrassment in this field, so that it ultimately operated rather as a check upon the progress of knowledge and understanding.

We have already seen the mechanical character of William Harvey's enquiries into the circulation of the blood; but at Harvey's university of Padua there is more remarkable direct evidence of the fact that the work and the principles of Galileo were having their effect on the medical faculty. Sanctus Sanctorius (1561-1636) set out to adapt the thermometer for clinical purposes, and used an instrument invented by Galileo for measuring the beat of the pulse. He studied temperature, respiration, the physics of the circulation, and in particular made experiments in weighing—he had a balance in which he could both eat and sleep and so could test his own weight under various conditions; and he made discoveries concerning insensible perspiration. Then there was a younger person, Giovanni

Alfonso Borelli (1608-79), who was a mathematician and a friend of Galileo. His book *On the Motion of Animals,* published in 1680-81, just after his death, represented a supreme example of the application of the science of mechanics to the study of the living organism. He was most successful in the treatment of muscular movements, using mathematics and diagrams, so that his treatise looks like a text-book in mechanics. One of his chapters deals with the "Mechanical propositions useful for the more exact determination of the motive power of muscles." He examined the act of walking as nobody had ever done before, and then turned to the flight of birds and the swimming of fish—almost the first thing he asks about a bird is where its centre of gravity lies. One of his sections is entitled: "The quantity of air acted upon by the wing of a Bird in flight is in shape a solid sector swept out by a radius equal to the span of the wing." He calculates that "the power of those muscles that beat the wings is greater by ten-thousand times than the weight of the bird," and he points out that, if anything analogous applies in the case of human beings, the motive power of our pectoral muscles could never be sufficient for such a task, so that the old story of Icarus could not possibly have been true. Starting from the work of William Harvey, he examined the action of the fibres of the heart, and calculated that, to maintain the circulation, the heart at each beat must exert a force equivalent to not less than 135,000 lb. We find him comparing the heart to a piston or a winepress. Also, he worked out that if the blood flows evenly from the arteries, through the minute capillaries, into the veins (for its return journey to the heart), this steady flow is due to the elastic reactions of the arterial walls. The arteries after expansion contract and force the blood forward as though a rope had been twisted round them—so that a certain regularity in the flow is not directly, but indirectly, attributable to the beats of the heart itself. A contemporary of Borelli's, the Dane, Niels Stensen, who worked chiefly in France and Italy, was also essentially a mechanist, seeking to apply mathematical and geometrical principles to the muscles. In proportion as the peculiarly mechanical approach was inadequate, in proportion as chemical operations were involved—as in the case of the digestion—this whole method of dealing with the living body was bound to prove a hindrance to biological studies sooner or later.

The result of this tendency to glorify mere mechanisation was the spread of the view that the animal body was nothing more than a piece of clockwork. Descartes, completing the continuous interlocking machinery of his physical universe, seems to have made himself the most remarkable exponent of this view. Adherence to his principle of strict animal automatism became, so to speak, a marginal point of dogma and was regarded as a test case amongst the followers of his system; it decided whether you could claim to have in you the pure milk of Cartesian orthodoxy. The whole issue was one which aroused great controversy in the seventeenth and eighteenth centuries. The philosophy of Descartes made so strict a separation between thought and matter, soul and body, that it was hardly possible to bridge the gulf between them with anything short of a miracle that defeated imagination. Under this system, animals were regarded as being devoid of either thought or genuine consciousness, whereas to possess precisely these two things was the very essence of the human soul. It was even held, therefore, that animals could not authentically see anything or feel the bitterness—the genuine pangs—of bodily pain. Their eyes would hold the kind of picture which we have when we are not ourselves—when we gaze on things with a glassy stare and a vacant look but do not really apprehend. Similarly, in the theory of Descartes, animals had a purely corporeal, unconscious kind of sensation, but no consciousness, no mental agony, no real ability to feel pain. God, the human soul and the whole realm of spiritual things, however, escaped imprisonment in the process of mechanisation, and were superadded presences, flitting vaporously amongst the cogwheels, the pulleys, the steel castings of a relentless world-machine. It was very difficult to show how these two planes of existence could ever have come to intersect, or at what point mind or soul could ever join up with matter. There was a sense in which the soul—representing chiefly Thought in the Cartesian system—could hardly be regarded as having a location in space at all. There was another sense in which it was possible to say that it was no more associated with one part of the human body than with another. Descartes particularly attached it, however, to the pineal gland, partly because it was thought that the mere animals did not possess this feature at all. In the next generation, however, Niels Stensen spoiled the argument by discovering this gland in other animals. Descartes, having imagined that he had found the strategic place where soul and body joined, acquired a certain virtue by concentrating attention at the next remove upon the action of the nerves. He believed, however, that an actual transmission of matter took place from the nerves to the muscles. In general, he arrived at too direct a process of mechanisation.

Leibnitz said that everything which took place in the body of man or animal was as mechanical as the things that happen inside a watch. Some Englishmen later in the seventeenth century—Henry More, the Cambridge Platonist, for example—thought that Descartes had gone too far in his idea of man as an organised statue, an automatic machine. Newton felt that, though the system of Descartes necessitated a Creator who had set the clockwork into motion in the first place, it was in danger of making God superfluous once the universe had been given a start. And it is curious to note that, if earlier in the century religious men had hankered after a mathematically interlocking universe to justify the rationality and self-consistency of God, before the end of the century their successors were beginning to be nervous because they saw the mechanism becoming possibly too self-complete. Boyle differed from Descartes in thinking that God was necessary, not merely to set things in motion and to establish the laws of motion, but also to combine the atoms or corpuscles into those remarkable architectural systems that enabled them to organise themselves into a living world. Newton was even prepared to believe that gravity, which was otherwise so apparently unaccountable, represented the constant activity of a living being that pervaded the whole of space. He was ready to think also that special combinations which took place on occasion in the sky—unusual conjunctures, for example, or the passage of comets along a path when another heavenly body happened to be near—produced small mechanical derangements and casual discrepancies which necessitated the occasional intervention of a watchful Deity.

The effects of the scientific revolution in general, and of the new mechanistic outlook in particular, are vividly illustrated in the works of the Hon. Robert Boyle, who lived from 1627 to 1691, and who at the same time helps to demonstrate the importance of some of the ideas of Francis Bacon. From the age of twenty he came under the influence of the members of a group who from 1645 were meeting in London to study the New Philosophy, specifically described by them as the Experimental Philosophy. At the age of twenty, writing to another member of this group, he confessed that he had once been very much inclined to Copernicus, but he now wrote of the Ptolemaic and Copernican systems, as well as the system of Tycho Brahe, as though they were rival theories in a controversy which at the moment it was impossible to settle. Five or six years after this—that is to say, in 1652-53—he appears to have been converted to the doctrine of the circulation of the blood, and he made the acquaintance of William Harvey at the end of Harvey's life. In 1654 he went to live in Oxford at the invitation of Dr. John Wilkins, the recently appointed Warden of Wadham College, around whom were grouped a number of chemists, physicians, etc., who were passionate believers in Bacon and the Experimental Philosophy. Although he talks of having discovered "the usefulness of speculative geometry for natural philosophy," Boyle regrets his lack of mathematics and his work wears a more Baconian appearance because he does not give that mathematical turn to his researches. In parts of his works he sets out explicitly to justify the non-mathematical approach to the problems of science.

Boyle sought in the first place to be an historian in Bacon's sense of the word—the sense that is implied in the term "natural history"—namely, to assemble the results of particular enquiries and to accumulate a great collection of data which would be of use in the future to any person wishing to reconstruct natural philosophy. In this sense he set out to produce such collections as a natural history of the air, a history of fluidity and firmness, an experimental history of colour or of cold, just as Bacon had done with winds or with heat. He tells us that one of his collections was designed as a continuation of *Sylva Sylvarum*—that is to say, of Francis Bacon's natural history. He was so Baconian, he confesses, that for a long time he declined to read Gassendi or Descartes or even Bacon's own *Novum Organum,* lest he should be seduced too early by lofty hypotheses; though he agrees that there is a place for hypotheses behind experiments, and only insists that these should be of a subordinate kind, and that the scientist shall not cling to them too long or build too great superstructures upon them. Occasionally he says he will communicate the results of experiments without any theories concerning their causes, because it is the concrete data established by his enquiries which will be of permanent use, whether his theories prove to be acceptable or not. He created a certain public interest in his experiments as he employed a host of "assistants, experimenters, secretaries and collectors."

It was noted by the contemporary scientist, Huygens, and has been repeated with some justice by historians of science since, that in relation to the vast amount of experimental work that he put on record, Boyle made very few important discoveries or strategic changes in science. He wrote like Bacon: "It has long seemed to me to be none of the least impediments of the real advancement of true natural philosophy that men have been so forward to write systems of it." He complained that even the new science, the mechanical philosophy, still had too narrow an experimental basis, and that people despised the limited generalisations of the Baconian experimentalist—the world wanted everything to be explained from first principles after the manner of Descartes, and thought it nothing if you merely proved that a certain phenomenon was the result of the application of heat. He valued Bacon even where Bacon has most been despised by modern writers—namely, in his natural history. And though Boyle had severe things to say about Bacon's contemporary—the famous chemist, Van Helmont—as well as about many other people, he found it hard to believe that the great Francis Bacon could have been foolish. In one or two cases where people jeered at Bacon for reporting experiments that had proved fallacious, he set out to enquire how Bacon could have been wrong, and discovered that, for example, Bacon was correct if you assume that he used a purer sort of spirits of wine than was usually employed a generation later. Even apart from this, he was interested in those anomalies or impurities which often existed in the materials employed by chemists, and which explain why so many of their experiments were vitiated. He wrote about the whole range of accidents which in given cases prevented experiments from producing the correct result or even a uniform result. He kept careful registers of observations himself and insisted on the importance of recording experiments, of confirming them by unwearied repetition, and of distrusting a great amount of what purported to be the published record of experiments. He was Baconian in that he repeatedly thanked God that he had been initiated into chemical operations by illiterate artisans—men not capable of infusing into his mind the notions and philosophising of the alchemists, a class of people who had been blinded by the jargon of their trade. Perhaps he was Baconian even in his interest in what is more popularly known as alchemy, his belief that he could turn water into earth, his idea that he had transmuted gold into a lower metal, and the degree of secrecy and mystification which he manifested in regard to some of his work. It was because of his confidence in his work on that side that he petitioned parliament in 1689 and secured the repeal of a law of Henry IV's reign against people who set out to multiply gold and silver.

It is difficult for us to imagine the state of chemical enquiry before the days of Boyle, or to comprehend on the one hand the mystifications and the mysticisms, on the other hand the anarchical condition of things, amongst the alchemists in general. Van Helmont, who stands roughly twenty years after Francis Bacon, made one or two significant chemical discoveries, but these are buried in so much fancifulness—including the view that all bodies can ultimately be resolved into water—that even twentieth-century commentators on Van Helmont are fabulous creatures themselves, and the strangest things in Bacon seem rationalistic and modern in comparison. Concerning alchemy it is more difficult to discover the actual state of things, in that the historians who specialise in this field seem sometimes to be under the wrath of God themselves; for, like those who write on the Bacon-Shakespeare controversy or on Spanish politics, they seem to become tinctured with the kind of lunacy they set out to describe.

Two things, however, are clear in the time of Boyle, because the conscious campaign against them was one of the explicit objects of much of his writing. On the one hand he quarrelled with the scholastic interpretation of the properties and qualities—greenness, fluidity, coldness, etc.—which bodies possess; that is to say, he quarrelled with a traditional doctrine of what were called "substantial forms," a doctrine which he claimed had become hardened and perverted since its exposition by Aristotle. In this connection he showed that the doctrine of "substantial forms" had explained nothing and had only added a species of mystification, while it was clear that, without the embarrassment of that doctrine, important problems had been solved in recent times in statics, hydrostatics, etc. On the other hand, in regard to the constitution of matter, the followers of Aristotle believed that substances could be resolved into the four elements—earth, water, air and fire. The alchemists, whom Boyle was in the habit of calling Spagyrists, believed that matter could be resolved into three hypostatical principles—sulphur, salt and mercury. In his attack upon these views or upon any combination of them, Boyle came near to laying the foundations of modern chemistry and made his significant contributions to science—contributions relating to the structure of matter. Here his work is so stimulating that some historical explanation has to be provided for the fact that it took another century to place the science of chemistry really on its feet.

At a point where his work was to be of such interest and importance in the history of science, it is curious to note that Robert Boyle, in spite of all his preaching against that kind of thing, and in spite of the way in which he apparently tried to fight against himself, was spurred and stimulated by a species of doctrine, an all-embracing philosophy which had become current amongst the more advanced thinkers of the time. In a sense he was Baconian even in this—for it was the corpuscular view of the universe which had attracted him, and Bacon in one of his most fascinating essays, as we have seen, had called attention to the corpuscular theory, making the significant remark that it was either true or useful for the purposes of demonstration, since hardly any other hypothesis enabled one either to comprehend or to portray the extraordinary subtlety of nature. Far from regarding it as a purely speculative theory of the kind that ought to be avoided, Bacon had told the scientists that this was precisely the direction in which they ought to move if they wished to "cut nature to the quick." Boyle tells us that for a long time he avoided reading about the corpuscularian hypothesis lest his work and his mind should be deflected by it, but it is clear from his writing that he could not really resist it. He continually discussed it whether in the form of an atomic theory in Gassendi, or in the system of Descartes, which regarded matter as infinitely divisible, and he bracketed the two as forms of the corpuscularian theory, setting them against both the Aristotelian and the alchemical theories of matter. He said on one occasion that he was prepared to be corrected in his particular generalisations about the formations of mixtures and compounds in chemistry, but that as a natural philosopher he did not expect to "see any principles proposed more comprehensive and intelligible than the corpuscularian," which he often also called the "mechanical" philosophy, since it purported to give a mechanical explanation of the physical universe. Here, then, is another way in which Boyle is a product of the scientific revolution on its essentially mechanical side.

Boyle said that the Peripatetics, the Aristotelians, made too little use of experiment—they brought in experiment merely to illustrate the principles they had arrived at in their general system of philosophy. That was a criticism very Baconian in character, and perhaps not quite just, and the same was true of the continued attacks upon the syllogistic method. Boyle was Baconian again when he pointed out that the mechanical philosophers themselves "have brought few experiments to verify their assertions." He consciously set out to do that particular service to the mechanical philosophers—to supply the experimental background for their theory of matter; and in that sense, in spite of the multitude of his protestations, he was laying himself open to the charge of doing the very thing he had reproached the Aristotelians for doing—namely, using experiments to demonstrate and fortify a philosophy already existing in his mind. Above all, Boyle tells us that he set out to "beget a good understanding between the chymists and the mechanical philosophers who have hitherto been too little acquainted with one another's learning." He stressed the necessity for an alliance between chemistry and mechanical science in the study of the body—stressed the importance of having men actually skilled in both sciences at once—and showed that chemistry had its part to play, for example in research on the subject of the digestion. He regarded the whole corpuscular philosophy as being confirmed by chemical science, in the operations of which, he said, it often happened that "matter [was] divided into parts too small to be singly sensible." On the one hand, therefore, the principles of mechanics are often brought by Boyle into connection with chemical and medical questions—he has a work, for example, entitled *Medicina Hydrostatica or Hydrostatics applied to Materia Medica.* On the other hand, he constantly hauls these sciences into connection with the fashionable seventeenth-century hypotheses concerning the structure of matter. One of his writings is entitled *Of the Reconcileableness of Specific Medicines to the Corpuscular Philoso-*

phy. He repeatedly expressed his anxiety to show that chemical experiments had relevance and applicability to natural philosophy in its higher reaches. In all that it is abundantly clear that he thought he was doing the greatest possible service to Christianity, for the interests of which he was very jealous and for the furtherance of which he wrote many treatises.

In his attacks both on Aristotle and on the alchemists, he gives a close picture of the structure of matter as conceived in the new mechanical philosophy, and we can follow the way in which he was led to his new doctrine concerning the chemical elements. On his view, the universe could be explained from three original principles—matter, motion and rest—matter itself being capable of reduction to minute particles which on one occasion and for purposes of illustration he was prepared to assume to be a billionth of an inch in length. Firstly, he said:

> there are in the world great stores of particles of matter, each of which is too small to be, while single, sensible; and being intire or undivided must needs have both its determinate shape and be very solid. Insomuch that though it might be mentally and by divine omnipotence divisible, yet by reason of its smallness and solidity nature doth scarce ever actually divide it; and these may in a sense be called *minima* or *prima naturalia.*

At the next stage, he said:

> there are also multitudes of corpuscles which are made up of the coalition of several of the former [particles or] *minima naturalia;* and whose bulk is so small and their adhesion so close and strict that each of these little primitive concretions or clusters . . . of particles is singly below the discernment of sense; and though not absolutely indivisible by nature into [the original particles or] *prima naturalia* that composed it . . . they very rarely happen to be actually

> dissolved or broken but remain intire in a great variety of sensible bodies.

Granted the original particles and their clustering together in knots or concretions, says Boyle, the purely mechanical movements and arrangements of these minute corpuscles will explain all the different characteristics and tendencies of physical bodies, so that there is no need to resort to Aristotelian notions of forms or to any mystification about the quality of greenness in bodies that happen to look green. The differences between one substance and another are merely differences in the schematic systems into which the particles of common matter are ranged, the motions that take place amongst them, and the difference of texture or structure which the various possible combinations produce. The configuration of the corpuscles, the size of the clusters, the position or posture of the particles—these are sufficient to explain all the variety that exists in nature. One of Boyle's works is entitled *Experiments about the Mechanical Origin or Production of Particular Qualities,* and it includes a discourse on the mechanical origin of heat and of magnetism. Boyle showed elsewhere that bodies are fluid when the minute corpuscles lie over one another, touching at only some parts of their surfaces, so that they easily glide along each other until they meet some resisting body "to whose internal surface they exquisitely accommodate themselves." "Bodies exhibit colours not upon the account of the predominancy of this or that principle in them but upon that of their texture and especially the disposition of their superficial parts; whereby the light rebounding thence to the eye is modified." Whiteness is the result of reflection from a body whose surface

> is asperated by almost innumerable small surfaces; which being of an almost specular nature [like tiny convex mirrors] are also so placed that some looking this way and some that, they yet reflect the rays of light that fall on them not towards one another but outwards towards the spectator's eye.

And just as the colours of plush or velvet will vary as you stroke one part of the fabric one way and another part another way—just as the wind creates waves of colour and shadow in a field of corn as it falls differently in different parts of it—so the posture and inclination of the particles in a given body will govern the way the light is modified before it is returned to the eye. Something similar to this is true with various processes in nature or in chemistry—in the case of putrefaction, for example, air or some other fluid fetches out the looser particles, and the substance is dislocated, producing perhaps even a change in the composition of the separate corpuscles. The taste of things is accounted for in a parallel way:

> If bodies be reduced into a multitude of parts minute and sharp enough, it is very possible that some of these, either in part or in conjunction with others, may acquire a size and shape that fits them sensibly to affect the organ of taste.

He speaks sometimes as though it were a case of their sharp edges pricking. His total view is clear—the qualities and properties of the bodies that we know may be accounted for by motion, size, configuration and combinations of particles. The behaviour of the particles and the resulting manifestations in the various kinds of body that exist are attributed by him to what he calls "the mechanical affections of matter," because, he says, they are analogous to "the various operations of mechanical engines." He often talks of the human body as a "matchless engine" and of the universe as "an automaton or self-moving engine."

He was greatly concerned over what he was for ever calling the texture or the structure of matter, the result of the numberless combinations that were possible with particles and corpuscles. Many of his greatest works are illustrations of his tremendous interest in this question; and it is not surprising that his most important contribution to chemistry lay in this field—namely, in his discussion of what constituted a chemical element. His most famous work, *The Sceptical Chymist,* addresses itself to this particular question. It does not give his positive system at its ripest—we must rather regard it as his greatest piece of destructive work. He attacked what his predecessors had hitherto regarded as the virtually irreducible things in chemistry—on the one hand, Aristotle's doctrine of the four elements; on the other hand, the teaching of the alchemist on the subject of the three hypostatical principles. He showed that the alchemists were wrong in assuming that by the use of fire all mixed bodies could be analysed into their elementary ingredients. He demonstrated that, indeed, different things happened if a sample of some mixed body were burned in an open fire and another sample in a closed retort. The results produced on a substance by a moderate degree of heat, he said, would not always be compatible with the effects produced by a very great heat. He showed that sometimes fire actually united bodies of different natures or produced out of a substance new compounds that had not existed previously. Having combined two ingredients to form soap, he heated the soap in a closed retort and produced out of it two substances different from the ones he had used originally. He maintained that fire divides mixed bodies because some parts are more fixed and others more volatile, but that it does not matter whether either of these is of an elementary nature or not—fire does not necessarily reduce a substance into its primary elements. At the same time he pointed out that nobody had ever divided gold into any four component parts, while blood was a substance capable of being reduced to more than four ingredients.

He drew attention to the difference between chemical compounds and mere mixtures, showed how the two were differently related to the elements composing them, and indicated the tests that would facilitate the identification of individual substances. In a full-dress attack on the prevailing belief in the three Paracelsan "principles"—sulphur, salt and mercury—he clarified the description of the irreducible nature of a chemical element, though Van Helmont had foreshadowed him in this, and indeed it can be argued that, through the way in which he apprehended the matter, Boyle introduced a new confusion which nullified the benefit he had bestowed. He showed that fire was unable to reduce glass to its elements, though everybody was aware that there were elements to which it could be reduced, since it was composed of sand and alkali. He regarded a piece of gold as being built up from very fine corpuscles and he was prepared to believe that these metallic corpuscles were even more resistant to sub-division than glass itself. He regarded them as composed of ultimate particles, each corpuscle being what he called a "concretion"—extremely stable, extremely difficult to reduce, and clearly recoverable even after gold had been compounded with something else and had apparently disappeared. But he was not convinced that they could never be resolved into something more genuinely elementary; and he does not seem to have been consistenly prepared to be pragmatic, as Lavoisier was ready to be at a later date, and to accept a substance as "elementary" merely because it had proved chemically irreducible up to this time. On occasion he even expressed doubts about the existence of "chemical elements" or the need for postulating such things, since the differences between one substance and another might be explained as the effect of size, shape, structure, texture, and motion produced by the mere accretion and architectural arrangement of the ultimate particles of primary matter. Chemistry itself could be reduced to what has been called "micro-mechanics," therefore; and Boyle himself showed a tendency to rush direct to this ultimate explanation of the qualities that he found in any kind of matter. In this, if to some degree he foreshadowed a distant future, he may still have done harm, because it meant by-passing the whole idea of a chemical element. His mechanistic philosophy may have helped him in some ways, but it hindered him in other ways, therefore. And, at this point in the argument his work, in proportion as it influenced the world, may well have served rather to check the progress of chemistry. The Paracelsan division of matter into three hypostatical "principles" may have gone somewhat out of fashion in the subsequent period; but the parallel system of Aristotle—the doctrine of the four "elements"—was to recover favour, as was noted in the eighteenth century.

Boyle's study of the atmosphere, which inaugurated his career as a chemist and his quarrel with Aristotelianism, has an important place in a longer narrative of seventeenth-century discovery. When Galileo was faced with two smooth slabs of marble or metal so clinging together that the one could lift the other, he interpreted the phenomenon in accordance with the Aristotelian thesis that nature abhors a vacuum, and he slurred over the objection that the resistance of these bodies to any attempt to separate them could hardly be due to something which did not yet exist—namely, the vacuum that *would* be produced by their separation. When a pump refused to raise water above thirty-two feet Galileo did not ask why nature's abhorrence of a vacuum should come to its limit at that point—he said that the column of water broke by its own weight, just as he said one might hang up a column of iron so long and heavy that it would break by its own weight. Galileo both regarded the atmosphere as having weight and conjectured that a column of mercury, because it was so much heavier than water, would break if it reached only one-fourteenth of the height that a pump could carry water. But it was his disciple, Torricelli, who took a tube three feet long (sealed at the top), filled it with mercury, and immersed it in a bowl of mercury, so that the column of liquid fell to two feet six inches; thereby showing that the pressure of the atmosphere kept up the column of mercury and that something like a vacuum existed at the top of the tube. This led to the discovery of the barometer and then to further experiments concerning atmospheric pressure—its variation at different altitudes, for example—while in Germany the observation of the pumping of water led to the important discovery of the air-pump.

Robert Boyle greatly improved the German air-pump which, he tells us, required the hard labour of two men for many hours before the vessel could be emptied. He demonstrated that the air could be weighed—that it had an expansive force which resisted pressure and that the barometric column was kept up solely by the weight of the outside air. At one moment he made the interesting conjecture that the behaviour of the air might be explained by regarding its minute particles as so many tiny coiled-up springs. In addition to all this, he studied both respiration and combustion and came near to the discovery of oxygen when he said: "There is in the air a little vital quintessence (if I may call it so) which serves to the refreshment and restoration of our vital spirits, for which uses the grosser and incomparably greater part of the air [is] unserviceable." He realised that in the atmosphere there is "a confused aggregate of effluviums. . . . There is scarce a more heterogeneous body in the world." But it seems to have been his view that the air itself was homogeneous, the variations being due to the presence of steams and effluvia which in reality were foreign to it. In this single field his experiments did much to justify the Baconian influence under which they were undertaken—the principle that the scientist should use the experimental method in order to collect concrete data, without attempts at too hurried synthesis. And if his concentration on the mechanical activity of the air may have had unfortunate effects on the chemical study of the atmosphere in the following period, as we shall see, Boyle in general marks so great a difference from the older chemistry that historians have had to wonder why greater progress was not made in that science in the next century.

Chapter 8

The History of the Modern Theory of Gravitation

IT BECAME CLEAR to us when we were studying the work of Copernicus that the hypothesis of the daily and annual rotation of the earth presented two enormous difficulties at the start. The first was a problem in dynamics. It was the question: What power was at work to keep this heavy and sluggish earth (as well as the rest of the heavenly bodies) in motion? The second was more complicated and requires some explanation; it was the problem of gravity. On the older theory of the cosmos all heavy bodies tended to fall to the centre of the earth, because this was the centre of the universe. It did not matter if such earthy and heavy material were located for a moment on the immaculate surface of a distant star—it would still be drawn, or rather would aspire to rush, to the same universal centre, the very middle of this earth. Indeed, supposing God had created other universes besides ours and a genuine piece of earthy material found itself in one of these, it would still tend to fall to the centre of our universe, because every urge within it would make it seek to come back to its true home. Granted an earth which described a spacious orbit around the sun, however, such a globe could no longer be regarded as the centre of the universe. In that case, how could the existence of gravity be explained? For it was still true that heavy objects seemed to aspire to reach the centre of the earth.

The two problems in question became more acute when, towards the end of the sixteenth century, men began to see the untenability of the view that the planets were kept in motion and held in their proper courses by their attachment to the great crystalline orbs that formed the series of rotating skies. It became necessary to find another reason why these heavenly bodies should keep in movement yet not drift at the mercy of chance in the ocean of boundless space. These two problems were the most critical issues of the seventeenth century, and were only solved in the grand synthesis produced by Sir Isaac Newton in his *Principia* in 1687—a synthesis which represented the culmination of the scientific revolution and established the basis of modern science. Though it entails a certain amount of recapitulation, we shall be drawing the threads of our whole story together if we try to mark out the chief stages in the development of this new system of the universe.

It has been suggested that Copernicus owes to Nicholas of Cusa his view that a sphere set in empty space would begin to turn without needing anything to move it. Francis Bacon said that before the problem of the heavens could be solved it would be necessary to study the question of what he called "spontaneous rotation." Galileo, who seems at times almost to have imagined gravity as an absolute—as a kind of "pull" which existed in the universe irrespective of any particular object in space—drew a fancy-picture of God dropping the planets vertically until they had accelerated themselves to the required speed, and then stopping the fall, turning it into circular motion at the achieved velocity—a motion which on his principle of inertia could then be presumed to continue indefinitely. Involved in the whole discussion concerning the form of the universe was the special problem of circular motion.

Copernicus was responsible for raising these great issues and, as we have already seen, he did not fail to realise the magnitude of the problems he had set. It was his view that other bodies besides the earth—the sun and the moon, for example—possessed the virtue of gravity; but he did not mean that the earth, the sun and the moon were united in a universal gravitational system or balanced against one another in a mutual harmony. He meant that any mundane object would aspire to regain contact with the earth, even if it had been carried to the surface of the moon. The sun, the moon and the earth, in fact, had their private systems, their exclusive brands, their appropriate types of gravity. For Copernicus, furthermore, gravity still remained a tendency or an aspiration in the alienated body, which rushed, so to speak, to join its mother—it was not a case of the earth exercising an actual "pull" on the estranged body. And, as we have already seen, Copernicus regarded gravity as an example of the disposition of matter to collect itself into a sphere. The Aristotelian theory had implied the converse of this—the earth became spherical because of the tendency of matter to congregate as near as possible around its centre.

In view of the principles which were emphasised in this way in the system of Copernicus, a special significance attaches to the famous book which William Gilbert published in 1600 on the subject of the magnet. This work marks, in fact, a new and important stage in the history of the whole problem which we are discussing. I have already mentioned how, according to Aristotle, four elements underlay all the forms of sublunary matter, and one of these was called "earth"—not the soil which we can take into our hands, but a more refined and sublimated substance free from the mixtures and impurities that characterise the common earth. William Gilbert, starting from this view, held that the matter on or near the surface of the globe was waste and sediment—a purely external wrapping like the skin and hair of an animal—especially as by exposure to the atmosphere and to the influence of the heavenly bodies it was peculiarly subject to debasement and to the operation of chance and change. The authentic "earth"—Aristotle's element in its pure state—was to be found below this superficial level, and formed, in fact, the bulk of the interior of this globe. Indeed, it was neither more nor less than lodestone. This world of ours was for the most part simply a colossal magnet.

The force of the magnetic attraction was the real cause of gravity, said Gilbert, and it explained why the various parts of the earth could be held together. The force of the attraction exerted was always proportional to the quantity or mass of the body exerting it—the greater the mass of the lodestone, the greater the "pull" which it exercises on the related object. At the same time, this attraction was not regarded as representing a force which could operate at a distance or across a vacuum—it was produced by a subtle exhalation or effluvium, said Gilbert. And the action was a reciprocal one; the earth and the moon both attracted and repelled one another, the earth having the greater effect because it was so greatly superior in mass. If a magnet were cut in two, the surfaces where the break had been made represented opposite poles and had a hankering to join up with one another again. Magnetism seemed to represent the tendency of parts to keep together in a whole, therefore—the tendency of bodies, of material units, to maintain their integrity. Gilbert's view of gravity carried with it an attack on the idea that any mere geometrical point—the actual centre of the universe, for example—could operate as the real attraction or could stand as the goal towards which an object moved. Aristotle had said that heavy bodies were attracted to the centre of the universe. The later scholastics who adopted the impetus-theory—Albert of Saxony, for example—had developed this view and had brought out the point that in reality it was the centre of gravity of a body which aspired to reach the centre of the universe. Gilbert, on the other hand, insisted that gravity was not an action taking place between mere mathematical points, but was a characteristic of the stuff itself, a feature of the actual particles which were affected by the relationship. What was important was the tendency on the part of matter to join matter. It was the real material of the magnet that was engaged in the process, as it exercised its influence on a kindred object.

Francis Bacon was attracted by this view of gravity, and it occurred to him that, if it was true, then a body taken down a well or mine—into the bowels of the earth—would perhaps weigh less than at the surface of the earth, since some of the attraction exerted from below would possibly be cancelled by magnetic counter-attraction from that part of the earth which was now above. And though there were fallacies in this hypothesis, the experiment was apparently attempted more than once in the latter half of the seventeenth century; Robert Hooke, for example, stating that he tried it on Bacon's suggestion, though he failed to reach a satisfactory result. Gilbert's views on the subject of gravity took their place amongst the prevailing ideas of the seventeenth century, though they did not remain unchallenged, and it was long confessed that the question presented a mystery. Robert Boyle wrote of gravity as being possibly due to what he called "magnetical steams" of the earth. He was prepared, however, to consider an alternative hypothesis—namely, that it was due to the pressure of matter—the air itself and the ethereal substances above the air—upon any body that happened to be underneath.

William Gilbert constructed a spherical magnet called a *terrella*, and its behaviour strengthened his belief that the magnet possesses the very properties of the globe on which we live—namely, attraction, polarity, the tendency to revolve, and the habit of "taking positions in the universe according to the law of the whole"—automatically finding its proper place in relation to the rest of the cosmos. Whatever moves naturally in nature, he said, is impelled by its own force and "by a consentient compact of other bodies"; there was a correspondence between the movement of one body and another so that they formed a kind of choir; he described the planets as each observing the career of the rest and all chiming in with one another's movements. That gravitational pull towards the centre affected not merely bodies on the earth, he said, but operated similarly with the sun, the moon, etc., and these also moved in circles for magnetic reasons. Magnetism, furthermore, was responsible for the rotation of the earth and the other heavenly bodies on their axes. And it was not difficult to achieve rotation even in the case of the earth, he said, because as the earth has a natural axis it is balanced in equilibrium—its parts have weight but the earth itself has no weight—it "is set in motion easily by the slightest cause." He held that the moon always turned the same face to the earth because it was bound to the earth magnetically. But, like Copernicus, he regarded the sun as the most powerful of all the heavenly bodies. The sun, he said, was the chief inciter of action in nature.

In a curious manner the wider theories of Gilbert had found the way prepared for them and had had their prospects somewhat facilitated. Since the fourteenth century there had existed a theory that some magnetic attraction exerted by the moon was responsible for the tides. Such an idea came to be unpopular amongst the followers of Copernicus, but it appealed to astrologers because it supported the view that the heavenly bodies could exercise an influence upon the earth. In the very year after the publication of Copernicus's great treatise—that is to say, in 1544—a work was produced which attributed the tides to the movement of the earth, and Galileo, as we have already seen, was to make this point one of his capital arguments in favour of the Copernican revolution. It was in reply to Galileo that the astrologer Morin put forward a view which had already appeared earlier in the century—namely, that not only the moon but also the sun contributed to affect the tides. Galileo at one time was prepared to adopt the more general theories of Gilbert in a vague kind of way, though he did not pretend that he had understood magnetism or the mode of its operation in the universe. He regretted that Gilbert had been so much a mere experimeter and had failed to mathematise magnetic phenomena in what we have seen to be the Galileian manner.

Even earlier than Galileo, however, the great astronomer Kepler had been influenced by Gilbert's book, and it appears that he had been interested somewhat in magnetism before the work of Gilbert had been published. Kepler must have an important place in the story because, under the influence of the magnetic theory, he turned the whole problem of gravity into a problem of what we call attraction. It was no longer a case of a body aspiring to reach the earth, but, rather, it was the earth which was to be regarded as drawing the body into its bosom. Put a bigger earth near to this one, said Kepler, and this earth of ours would acquire weight in relation to the bigger one and tend to fall into it, as a stone falls on to the ground. And, as in Gilbert's case, it was not now a mathematical point, not the centre of the earth, that exercised the attraction, but matter itself and every particle of matter. If the earth were a sphere, the stone would tend to move towards its centre for that reason, but if the earth were differently shaped—if one of its surfaces were an irregular quadrilateral, for example—the stone would move towards different points according as it approached the earth from one side or another. Kepler further showed that the attraction between bodies was mutual—the stone attracts the earth as well as the earth a stone—and if there were nothing to interfere with the direct operation of gravity, then the earth and the moon would approach one another and meet at an intermediate point—the earth covering one-fifty-fourth of the distance (assuming it to be of the same density as the moon) because it was fifty-four times as big as the moon. It was their motion in their orbits which prevented the earth and its satellite from coming into collision with one another in this way.

In Kepler we see that curious rapprochement between gravity and magnetism which was already visible in Gilbert, whom he admired so greatly, and which is explicit in later seventeenth-century writers. As in the example of the broken magnet this gravity could be described as a tendency in cognate bodies to unite. Kepler belongs also to the line of writers who believed that the tides were caused by the magnetic action of the moon; and he has been criticised on the ground that his chains of magnetic attraction, which he pictured as streaming out of the earth, were so strong as to have made it impossible to hurl a projectile across them. He did not quite reach the idea of universal gravitation, however—for example, he did not regard the fixed stars as being terrestrial bodies by nature and as having gravity, though he knew that Jupiter threw shadows and Venus had no light on the side away from the sun. Like Bacon, he seems to have regarded the skies as becoming more æthereal—more unlike the earth—as they receded from our globe and as one approached the region of the fixed stars. Also, he regarded the sun as a special case, with so to speak, a gravity of its own.

Having noted that the speed of planets decreased as the planet became more distant from the sun, he regarded this as a confirmation of the view to which he was mystically attached in any case—namely, that the sun was responsible for all the motion in the heavens, though it acted by a kind of power which diminished as it operated at a farther range. He held that the planets were moved on their course by a sort of virtue which streamed out of the sun—a force which moved round as the sun itself rotated and which operated, so to speak, tangentially on the planet. He once called this force an *effluvium magneticum* and seemed to regard it as something which was transmitted along with the rays of light. If the sun did not rotate, he said, the earth could not revolve around it, and if the earth did not rotate on its axis, the moon in turn would not revolve around our globe. The rotation of the earth on its axis was largely caused by a force inherent in the earth, said Kepler, but the sun did something also to assist this movement. Granting that the earth rotates 365 times in the course of the year, he thought that the sun was responsible for five of these.

Kepler knew nothing of the modern doctrine of inertia which assumes that bodies will keep in motion until something intervenes to stop them or to deflect their course. In his theory the planets required a positive force to push them around the sky and to keep them in motion. He had to explain why the motion was elliptical instead of circular and for this purpose he made further use of magnetism—the axis of the planets, like that of the earth, always remained in one direction, and at a given angle, so that now the sun drew these bodies in, now it pushed them away, producing therefore an elliptical orbit. The force with which the planets were propelled, however, did not radiate in all directions and distribute itself indiscriminately throughout the universe like light, but moved from the sun only along the plane of the planet's orbit. The force had to know, so to speak, where to find its object, therefore—not ranging over the whole void but aiming its shafts within the limits of a given field. In a similar way, the idea that the attracting body must be sensible of its object—the earth must know where the moon was located in order to direct its "pull" to that region—was one of the obstacles to the theory of an attraction exerted by bodies on one another across empty space.

The world, then, seemed to be making a remarkable approach to the modern view of gravitation in the days of Kepler and Galileo, and many of the ingredients of the modern doctrine were already there. At this point in the story, however, an important diversion occurs, and it was to have an extremely distracting effect even long after the time of Sir Isaac Newton himself. René Descartes—who, as we have seen, had undertaken, so to speak, to reconstruct the universe, starting with only matter and motion, and working deductively—produced a world-system which it is easy for us to underestimate today, unless we remember the influence that it had even on great scientists for the rest of the century and still later. It is only in retrospect and perhaps through optical illusions that—as in the case of more ancient attempts to create world-systems—we may be tempted to feel at this point that the human mind, seeking too wide a synthesis and grasping it too quickly, may work to brilliant effect, yet only in order to produce future obstructions for itself.

We have already seen that, in spite of all his attempts to throw overboard the prejudices of the past, Descartes was liable to be misled by too easy an acceptance of data that had been handed down by scholastic writers. It is curious to note similarly that two grand Aristotelian principles helped to condition the form of the universe as he reconstructed it—first, the view that a vacuum is impossible, and secondly, the view that objects could only influence one another if they actually touched—there could be no such thing as attraction, no such thing as action at a distance. As a result of this, Descartes insisted that every fraction of space should be fully occupied all the time by continuous matter—matter which was regarded as infinitely divisible. The particles were supposed to be packed so tightly that one of them could not move without communicating the commotion to the rest. This matter formed whirlpools in the skies, and it was because the planets were caught each in its own whirlpool that they were carried round like pieces of straw—driven by the matter with which they were in actual contact—and at the same time were kept in their proper places in the sky. It was because they were all similarly caught in a larger whirlpool, which had the sun as its centre, that they (and their particular whirlpools) were carried along, across the sky, so that they described their large orbits around the sun. Gravity itself was the result of these whirlpools of invisible matter which had the effect of sucking things down towards their own centre. The mathematical principles governing the whirlpool were too difficult to allow any great precision at this time in such a picture of the machinery of the universe. The followers of Descartes laid themselves open to the charge that they reconstructed the system of things too largely by deduction and insisted on phenomena which they regarded as logically necessary but for which they could bring no actual evidence. In the time of Newton the system of Descartes and the theory of vortices or whirlpools proved to be vulnerable to both mathematical and experimental attack.

At the same time certain believers in the *plenum*—in the Cartesian idea of a space entirely filled with matter—contributed further ingredients to what was to become the Newtonian synthesis. Descartes himself achieved the modern formulation of the law of inertia—the view that motion continues in a straight line until interrupted by something—working it out by a natural deduction from his theory of the conservation of momentum, his theory that the amount of motion in the universe always remains the same. It was he rather than Galileo who fully grasped this principle of inertia and formulated it in all its clarity. A contemporary of his, Roberval, first enunciated the theory of universal gravitation—applying it to matter everywhere—though he did not discover any law regarding the variation in the strength of this gravitational force as it operated at various distances. He saw a tendency throughout the whole of matter to cohere and come together; and on his view the moon would have fallen into the earth if it had not been for the thickness of the ether within the intervening space—the fact that the matter existing between the earth and the moon put up a resistance which counterbalanced the effect of gravity.

This was in 1643. It was in 1665 that the next important step was taken, when Alphonse Borelli, though he followed Kepler in the view that it needed a force emanating from the sun to push the planets around in their orbits, said that the planets would fall into the sun by a "natural instinct" to approach the central body if the effect of gravity were not counterbalanced by a centrifugal tendency—the tendency of the planets to leave the curve of their orbits, like a stone seeking to leave the sling. So, though he came short in that he failed to see the planets moving by their own inertial motion and failed to understand the nature of that gravity which drew the planets towards the sun, Borelli did present the picture of the planets balanced between two opposing forces—one which tended to make them fall into the sun, and another which tended to make them fly off at a tangent. In the ancient world—in a work of Plutarch's which was familiar to Kepler, for example—the moon had been compared to a stone in a sling, in the sense that its circular motion overcame the effect of gravity. Borelli was unable to carry his whole hypothesis beyond the stage of vague conjecture, however, because he failed to understand the mathematics of centrifugal force.

By this date (1665), most of the ingredients of Newton's gravitational theory were in existence, though scattered in the writings of different scientists in such a way that no man held them in combination. The modern doctrine of inertia had been put forward by Descartes and was quickly gaining acceptance, though people like Borelli, who has just been mentioned, still seemed to think that they had to provide a force actually pushing the planets along their orbit. The view that gravity was universal, operating between all bodies, had also been put forward, and on this view it became comprehensible that the sun should have a pull on the planets, and the earth should keep the moon from flying off into space. Now, in 1665, there was the suggestion that this gravitational movement was counterbalanced by a centrifugal force—a tendency of the planets to go off at a tangent and slip out of the sling that held them. All these ideas—inertia, gravitation and centrifugal force—are matters of terrestrial mechanics; they represented precisely those points of dynamics which had to be grappled with and understood before the movements of the planets and the whole problem of the skies were settled. But if you had these on the one hand, there were the findings of astronomers on the other hand, which had to be incorporated in the final synthesis—and these included Kepler's three laws of planetary motion; the one which described the orbits as elliptical; the one which said that a line between the sun and any planet covered equal areas in equal times; and the one which said that the square of the time of the orbit was proportional to the cube of the mean distance from the sun. It had to be shown mathematically that the planets would behave in the way Kepler said they behaved, supposing their motions were governed by the mechanical laws which I have mentioned.

Huygens worked out the necessary mathematics of centrifugal action, especially the calculation of the force that was required to hold the stone in the sling and prevent it rushing off at a tangent. He seems to have arrived at this formula in 1659, but he only published his results in this field in 1673, as an appendix to his work on the pendulum clock. It seems, however, that it never occurred to Huygens to apply his views of circular motion and centrifugal force to the planets themselves—that is to say, to the problem of the skies; and he seems to have been hampered at this point of the argument by the influence of the ideas of Descartes on the subject of the heavenly bodies. In 1669 he tried to explain gravity as the sucking effect of those whirlpools of matter with which Descartes had filled the whole of space, and he illustrated this by rotating a bowl of water and showing how heavy particles in the water moved towards the centre as the rotation slowed down. He also believed at this time that circular motion was natural and fundamental—not a thing requiring to be specially explained—and that rectilinear motion in the case of falling bodies, as with the particles in the rotating bowl of water, was, so to speak, a by-product of circular motion.

A writer on Keats has attempted to show how in the period before the production of the sonnet "On Reading Chapman's Homer" the poet had been setting his mind at play—and gradually making himself at home—in what might be described as the field of its effective imagery. Now there had been an experiment in terms of astronomical discovery, now an attempt to squeeze a poetical phrase from the experience of the explorer; but one after another had misfired. The mind of the poet, however, had traversed and re-traversed the field, and in the long run apparently a certain high pressure had been generated, so that when the exalted moment came—that is to say, when Chapman's *Homer* had provided the stimulus—the happy images from those identical fields rapidly precipitated themselves in the mind of the poet. The sonnet came from the pen without effort, without any apparent preparation, but, in fact, a subterranean labour had long been taking place.

So, as the seventeenth century proceeded, the minds of men had traversed and re-traversed the fields which we have been studying, putting things together this way and that, but never quite succeeding, though a certain high pressure was clearly being generated. One man might have grasped a strategic piece in the puzzle and, in a realm which at the time hardly seemed relevant, another scientist would have seized upon another piece, but neither had quite realised that if the two were put together they would be complementary. Already the scattered parts of the problem were beginning to converge, however, and the situation had become so ripe that one youth who made a comprehensive survey of the field and possessed great elasticity of mind, could shake the pieces into the proper pattern with the help of a few intuitions. These intuitions, indeed, were to be so simple in character that, once they had been achieved, any man might well ask himself why such matters had ever given any difficulty to the world.

The role of Newton in the story has been the subject of recent controversy, and doubt has even been thrown upon his claim to have made the effective synthesis while still a very young man. It has been shown that, up to a period little short of this, his notes give no evidence of any exceptional ability; but by 1665-66 he was making important discoveries both in optics and in mathematics, while in respect of the gravitational theory his own retrospective account of his discoveries is not to be lightly dismissed. It would seem then that, acting independently, he had found the required formulæ relating to centrifugal force in 1665-66, before the work of Huygens on this subject had been published. He had also discovered that the planets would move in something like conformity with Kepler's laws if they were drawn towards the sun by a force which varied in inverse proportion to the square of their distance from the sun—in other words, he had succeeded in giving mathematical expression to the operation of the force of gravity. On the basis of these results he compared the force required to keep a stone in a sling or the moon in its orbit with the effect of gravity (that is to say, with the behaviour of falling bodies at the earth's surface). He found that the two corresponded if one made allowance for the fact that gravity varied inversely as the square of the distance. He treated the moon as though it had been a projectile tending to rush off in a straight line but pulled into a curve by the effect of the earth's gravity; and he found that the hypotheses fitted in with the theory that the force of universal gravitation varied inversely as the square of the distance. The fall which the moon had to make (as a result of the earth's drag) every second, if it was to keep its circular path, bore the requisite proportion to the descent of the body falling here at the surface of the earth. The story of Newton and the apple is historical and was bound to have at least a sort of typical validity—for if it was not an apple it had to be some other terrestrial falling body that served as the basis for comparison. And the essential feat was the demonstration that when the new science of terrestrial mechanics was applied to the heavenly bodies the mathematics came out correctly. Newton would seem, therefore, to have achieved his essential synthesis in 1665-66, though he was dissatisfied with certain points in the demonstrations and put the work away for many years.

In the middle of the 1660s, Borelli, Newton, Huygens and Hooke were wrestling with various parts of the same planetary problems, some of them also treading on one another's heels in the study of the nature of light. Huygens had visited London, produced experiments for the Royal Society, corresponded with members of that Society, and tried out in England his pendulum clocks, which date back to 1657. But in England, experiments with the pendulum had started independently, and Christopher Wren, William Croone, William Balle and Laurence Rooke appear to have inaugurated the enquiry into laws of motion, Robert Hooke performing most of the experiments. It is hardly possible to discover the influence that Huygens and the English scientists must have had on one another in the development of this work. The 1670s must represent one of the greatest decades in the scientific revolution, if not the climax of the whole movement; and in both London and Paris there were circles of scientific workers whose achievements at this time were of a remarkable nature. So far as the gravitational theory is concerned, it is difficult to resist the view that in this period our attention ought to be directed not merely to Newton as an individual but to the combined operations of the English group. The Royal Society, we are told, following Baconian principles, sought to collect from all the world the data necessary for the establishment of the Copernican hypothesis; and, perhaps ideally at least, its members were assisting one another, "freely communicating their methods and pooling their gains." Here the names that are in the forefront are those of Isaac Newton, Robert Hooke, Edmond Halley and Christopher Wren. Hooke is amazing in the number, the variety and the ingenuity of his experiments as well as for his extraordinary fertility in hypotheses. He followed

Bacon in his attempt to demonstrate that the effects of gravity on a body must diminish as the body was sunk into the bowels of the earth. He sought to discover how far the effects were altered at great heights or in the region of the equator; and he threw light on the problem by observations and experiments on the pendulum. From the globular shapes of the heavenly bodies and the stable conformation of the ridges on the moon he deduced that the moon and the planets had gravity; and by 1666 he saw the motion of a comet (for example) as incurvated by the pull of the sun upon it, and suggested that the motion of the planets might be explicable on the kind of principles that account for the motion of a pendulum. In 1674 he was suggesting that by this route one could arrive at a mechanical system of the planets which would be "the true perfection of astronomy." He pointed out that, apart from the influence which the sun exerted on the planets, account had to be taken of the force which all the heavenly bodies must be presumed to be exerting on one another. By 1678 he had formulated the idea of gravitation as a universal principle; and by 1679 he, too, had discovered that the diminution of the force of gravity is proportional to the square of the distance. In this period it would appear that Newton put on to paper some remarks which might suggest that he was unsure about his own earlier theory; for the moment he seemed to be less firm in his apprehension than Hooke. At the same time he was called upon for mathematical help on various occasions. Some doubt seems to have been entertained (especially amongst Englishmen) on the subject of Kepler's law relating to the elliptical orbit of the planets. Newton provided a demonstration of the fact that the attraction exerted on the planets made it necessary to adopt the elliptical rather than the circular hypothesis. Hooke was to claim the priority in respect of the whole gravitational theory, and because Newton had been secretive about his work in 1665-66, because many of Hooke's own papers disappeared, and because Newton's memory was defective sometimes—or his accounts unreliable—the controversy on this subject has been renewed in recent years. But, apart from the evidence of Newton's earlier interest at least in the problem, Hooke did not produce the mathematical demonstration of his system. It can simply be said on his behalf that in the crucial period he was developing his mathematical powers more than was once imagined to be the case. His reputation has risen, with the development of historical research, therefore; though the glory of Newton has not been eclipsed.

It should be noted that, whereas Kepler had seen the planets as subject to forces which emanated from the sun, the view which Hooke had expounded and which Newton was to develop presented a much more complicated sky—a harmonious system in which the heavenly bodies all contributed to govern one another in a greater or lesser degree. The satellites of Jupiter leaned or reacted on one another as well as influencing the planet itself, while Jupiter in turn had a still more powerful hold upon them. The planet, however, together with its collection of satellites, was in the grip of the sun (upon which it exerted its own small degree of attraction), and was also within range of the influence of neighbouring planets. As Newton remarked later, the sun was so preponderant amongst these bodies that the influence of the smaller ones mattered little and, similarly, one might make small account of the influence of the moon upon the earth. At the same time it had been noticed, especially in England, that when Jupiter and Saturn came into closest proximity with one another their movements showed an irregularity which was never observed at any other point in the course of their travels. Also the moon caused a slight alteration in the earth's orbit. By virtue of similar perturbations in the planet Uranus in 1846 astronomers were able to deduce the existence of still another planet—Neptune—before the planet had actually been observed. The whole system was therefore much more complicated in the 1670s than had been envisaged in the early part of the century—the whole sky presented a more intricate set of mathematical harmonies. It was to be the virtue of the new theory of the skies that it explained some of the minor anomalies, and embraced a world of interactions much more comprehensive than anything which Kepler had envisaged.

It was in the middle of the 1680s that Isaac Newton returned to the problem. His greatest difficulty had apparently been due to the fact that though gravity operates, as we have seen, between all particles of matter, he had to make his calculations from one mathematical point to another—from the centre of the moon to the centre of the earth, for example. In 1685 he was able to prove, however, that it was mathematically correct to act upon this assumption—as though the whole mass of the moon were concentrated at its centre, so that the whole of its gravity could be regarded as operating from that point. It happened furthermore that though in the middle '60s the data upon which he worked may not have been radically wrong, still in 1684 he was able to make use of more accurate observations and calculations; for in 1672 a French expedition under Jean Picard had enabled simultaneous measurements of the altitude of Mars to be taken in Cayenne and in Paris, and the results of the expedition made it possible to secure a more accurate estimate of the sun's mean distance from the earth—which was worked out at 87 million miles, coming nearer to the modern calculation of 92 million—as well as revealing still more vividly the magnitude of the solar system. It was even possible now to have more accurate measurements of the dimensions of the earth itself. The achievements of this expedition, although they had found their way into print at an earlier date, only became widely known after a publication of 1684, and these materials were the ones employed by Newton when he made his final calculations and produced his system. In the middle '80s, therefore, there were converging reasons for his return to the problem he had been dealing with twenty years before; and this time he was satisfied with his results and demonstrations, which were completed in 1686 and communicated to the world in the *Principia* in 1687.

One of Newton's objects when he promulgated his system was to show the impossibility of that theory of vortices or whirlpools which Descartes had formulated. He showed that mathematically a whirlpool would not behave in the way that Descartes had assumed—a planet caught in a whirlpool would not act in conformity with Kepler's observations on the subject of planetary motion. Furthermore, it would not be possible for a comet to cut a straight path across the whole system, from one whirlpool to another, in the way that the theory required. In any case, if the whole of space were full of matter dense enough to carry round the planets in its whirlings, the strength of so strong a resisting-medium would have the effect of slowing down all the movements in the universe. On the other hand, it appears that even mathematicians did not immediately grasp the meaning and the importance of the *Principia,* and many people—especially those who were under the influence of Descartes—regarded Newton as unscientific in that he brought back on to the stage two things which had been driven out as superstitious—namely, the idea of a vacuum and the idea of an influence which could operate across space between bodies that did not touch one another. His "attraction" was sometimes regarded as a lapse into the old heresies which had attributed something like occult properties to matter. Actually he denied that he had committed himself to any explanation of gravity, or to anything more than a mathematical description of the relations which had been found to exist between bodies of matter. At one moment, however, he seemed privately to favour the view that the cause of gravity was in the ether (which became less dense at or near the earth, and least dense of all at or near the sun), gravity representing the tendency of all bodies to move to the place where the ether was rarer. At another time he seemed to think that this gravitation of his represented an effect that had to be produced by God throughout the whole of space —something that made the existence of God logically necessary and rescued the universe from the overmechanisation that Descartes had achieved. And, as we have seen, Newton believed also that certain irregular phenomena in the skies—rare combinations and conjunctures, or the passage of a comet—were liable to cause a

Chapter 9

The Transition to the *Philosophe* Movement in the Reign of Louis XIV

WE HAVE NOW glanced at what seem to be the strategic lines in the story of the seventeenth-century scientific revolution, and we have seen how that revolution was particularly connected with the study of motion both on the earth and in the sky, the story reaching its culmination in that synthesis of astronomy and mechanics which was achieved in the system of Sir Isaac Newton. The point at which we have now arrived must stand as one of the great moments in the history of human experience, because though problems were not completely solved and even Newton could not say what were the causes of gravitation, still it emerges from the whole narrative that here was one of those occasions when, by solving this problem and that, men acquired new habits of mind, new methods of enquiry—almost incidentally, so to speak, they founded modern science. Furthermore, they discovered in the system of Newton that earth and sky could be comprised in a single survey and reduced to one fundamental system of law, and this tended to alter their attitude to the whole universe. We have already seen that conscious attempts were made to extend the mechanistic system itself, as well as the scientific methods which had given such great results in physics, to cover also the case of chemical phenomena and to embrace even the field of biology. We have seen also that in conscious correspondence with this mechanistic system ancient atomic philosophies had been resurrected or were being moulded into new forms. It is not often that the historian can sweep into a single cluster such a wide range of intellectual changes, forming altogether such a general transformation of man's outlook.

All this, however, represents only a small corner in that vast range of significance which the scientific revolution possesses; and we should be wrong if now we did not turn aside for a moment to study the repercussions of the new thought upon the life and society of the seventeenth century. The history of science ought not merely to exist by itself in a separate pocket, and if we have isolated certain aspects of it and put the microscope on these, we have done so on the view that at this point in the narrative the intellectual changes had significance for general history in its broadest sense. It would be useful now, therefore, if we were to go back to the number we first thought of and see where the scientific movement of the seventeenth century finds its place in the total story of civilisation. In this connection we can hardly do better than take up the story precisely at the point which we have actually reached in our account of the scientific revolution—the time when the movement may be said really to have come into its own. The galvanic period even for our present purposes seems to lie in that decade of the 1680s when, as we have already seen, a handful of scientists in both London and Paris were crowning the story with remarkable achievements; in fact, the years which came to their peak in 1687 with the publication of Newton's *Principia.*

In order to understand the developments that took place we cannot begin better than by putting our fingers on the principal agent in the transmission of the results of the scientific movement to the outside world in this period—namely, the French writer Fontenelle. He is the most important single link between the scientific revolution and the *philosophe* movement. He has a special interest for us because he lived from 1657 to 1757 and spanned the great transition with which we are now concerned. And he is instructive because on the one hand he is in a sense the first of the French *philosophes;* while on the other hand he invented and exploited a whole technique of popularisation. He was secretary of the *Académie des Sciences* from 1699 to 1741, and it may be useful first of all to take some evidence which he supplies in his famous *Éloges*—the funeral orations on a great number of the scientists of this period, which he gave in his capacity as secretary of the Academy. After examining what is a critical point in the intellectual transition we shall later try to find the place of the whole story in a wider survey of the history of civilisation, so as to bring out some further aspects of the passage to what we usually call the age of reason.

If we were grading forms of historical evidence, most of us would be inclined to put funeral orations into the lowest and least trustworthy class of all. But it often happens in the case of any class of document that a witness is most important in the things that he was not intending to give away, and the historian is in the position of a detective—there is not anything in the world which may not provide him with a clue. Fontenelle, in fact, was an extremely subtle and diplomatic narrator, and even in what was supposed to be eulogy he could manage without offence to draw attention to the weaknesses of this scientist or that—as in the case of one man who was unduly jealous both of his colleagues and of inferiors. At the same time he seems to have used the orations in order to conduct a kind of propaganda on behalf of the scientific movement, and interesting things emerge from the character of that propaganda—for example, where he tilts against prevailing educational methods or against religious prejudice there is a considerable amount of information concerning the controversies of the time.

It is, however, by collating a great number of his short biographies and comparing the narratives that we discover some of the most interesting of the sidelights which Fontenelle was particularly well situated to provide. There is always a stage in the historiography of a movement, or a revolution, or a war, which might be described as the "heroic age"—the primitive period in the writing-up of the story, when people make myths, count their trophies and exult in the overthrow of the enemy, or, in the spirit of the Orangemen, hold commemorative dinners. There is a sense in which the *Éloges* of Fontenelle may be said to contribute something to our picture because they provide us with just this saga of the scientific revolution. If we examine some illustrations, it is not the names of particular scientists that matter or the specific things said about them. What concerns us, rather, is the pattern adopted in these biographies and the cumulative effect of the series as a whole.

It is clear from the character of this collection of lives that we are now dealing not with lonely pioneers like Galileo, but with a movement that is on the way to becoming generalised. It would appear from Fontenelle's biographies that those who joined the movement often sprang from bourgeois families, and remarkably often they are the sons of *avocats,* as he himself happened to be. Repeatedly it appears that they had originally been intended for the Church and that their fathers had insisted on giving them an education in theology. But over and over again the same pattern occurs in these biographies—the youth finds the prevailing educational methods irksome, finds that he is being given an education in mere words and not in real things. And then, on repeated occasions, the identical fairy-story seems to take place.

Of one person, Bernoulli, we are told that he saw some geometrical figures by chance and immediately responded to their charms—proceeding later to a study of the philosophy of Descartes. Another, Amontons, found himself deaf after an illness and had to cut short his formal education—so now he could turn his mind to whatever he pleased, and he began to study machines and dedicated himself to a design for the production of perpetual motion. Régis was intended for the Church, and, while he was getting tired of the length of time which he had to spend on an unimportant line of work, he chanced upon the Cartesian philosophy and immediately was struck by it. Tournefort discovered the philosophy of Descartes in his father's library and recognised straight away that this was the thing his mind had been looking for. Louis Carré was the same—he was to have been a priest, but the prospect was disgusting to him; then he discovered the philosophy of Descartes which opened up a new universe for him. Malebranche was so transported on reading Descartes, that he gave up everything else for the sake of the study of his philosophy. Varignon picked up a volume of Euclid by chance, and, charmed to see the contrast between this and the sophistries and obscurities that he had been taught in the schools, he allowed himself to be led by geometry to the reading of Descartes who came to him as a new light. The Dutchman, Boerhaave, was training for theology, but having tumbled upon geometry he could not resist it because of its invincible charm. It is all like the Christians recounting conversions in the early stages of a religious movement, when one man after another sees the light and changes the course of his whole life. And the movement generalises itself in those people who represent a new generation, glad to be emancipated from the burden or routine or prejudice of the old one. Everything is still in the heroic period—the scientific view comes as a new revelation and its apostles are counting their conquests. And a particular agent in the transition is geometry, especially the influence of Descartes, which was very powerful at this time.

Fontenelle's biographical sketches, furthermore, give evidence concerning the curious social success of the sciences in the reign of Louis XIV. As we survey them we find that the period of the 1680s comes out with peculiar vividness, especially as at this point he speaks very often as an eyewitness, and gives us what are really glimpses into his autobiography. It was probably from about 1680 that he himself was meeting some of the leading members of the new movement, and by 1683 or 1684 he had established himself in definite contact with it. This means that he came into direct touch with the scientists at an exhilarating moment—at one of the peak points in the story—and it is remarkable that though at this time he was only in the middle twenties, and had over seventy years longer to live, not only did his most vivid impressions belong to this period, but much of his permanent outlook was established by this date. We learn that Paris was full of foreigners who came to attend the conferences or observe the demonstrations of various scientists—in one year there were as many as forty people from Scotland alone who came to hear the famous chemist Lémery. Crowds of women attended Lémery's courses—carried away by the prevailing fashion, we are told—and as a result of his popularity his medicinal preparations became the vogue. Régis, the philosopher, had provided a stimulus in Toulouse, where he had awakened in both the clergy and the magistrates an interest in Cartesianism; and in 1680 he had come to Paris, where "the concourse of people was great—so great that a private house could not conveniently hold them; people came to make sure of seats long before the time to begin." The most pleasant of the actors at the Italian theatre came to learn the philosophy of Descartes; but the impression produced by Régis was so sensational that the Archbishop of Paris put an end to the sessions. Another man, Du Verney the anatomist, attracted young ladies to the study of his subject. "Anatomy, hitherto confined to the schools of medicine," writes Fontenelle, "now dared to come out into the *beau monde.*" He adds: "I remember having seen people of the *beau monde* carrying away the dried specimens which he had prepared in order to show them in their own circles." Later, when he was Professor at the *Jardin Royal,* this man Du Verney had great crowds of students, 140 foreigners attending his courses in a single year. We learn more about a group of men from Normandy who formed an interesting circle in Paris, including Fontenelle himself, Varignon the famous geometer, the well-known Abbé de Saint Pierre, and Vertot an historian. We learn of aristocratic patronage—how Lémery the chemist came to be admitted to the *salons* of the Prince de Condé, where many learned men assembled; while Du Verney was asked to assist in the education of the Dauphin, and then had a more distinguished audience than ever at his conferences. Fontenelle makes the interesting note that mathematics made a good profit in the 1680s, though he adds the significant remark that the outbreak of the great war in 1688—the war between Louis XIV and William III—had a bad effect on this.

Then in 1686 Fontenelle produced his famous dialogue entitled *The Plurality of Worlds*—the first work in France which made the discoveries of science clear, intelligible and amusing to the general reading-public. In many respects this book would stand as a model for works of scientific popularisation that have appeared down to the present day. Fontenelle consciously set out to make science amusing to fashionable ladies and as easy as the latest novel, and in this connection it is important to note that he had begun life as a writer before he ever turned to science—he was a literary man *manqué*—having written undistinguished verse and having failed as a dramatist. In one sense he was typical of the whole French *philosophe* movement of the eighteenth century which he helped to inaugurate. The results of the scientific revolution were precipitately and hastily translated into a new world-view, and this work was carried out not so much by scientists as by men of letters. Fontenelle wrote some mathematical works, but he was not himself important as a scientist—he is not remembered by any actual scientific discovery of his own. He was the proper person to write the funeral orations because he was somewhat versatile and familiarised himself with many sciences, so that he could appraise the discoveries of specialists in widely different fields. As a man of the world he saw what was fashionable and produced just what was wanted—bringing out his work on *The Plurality of Worlds* in the year before the publication of the *Principia,* and giving the general picture of the heavens as this had been achieved in the pre-Newtonian period. He invented, and carried to extremes, a playful style, based on a conversational manner, full of what was called *esprit,* with devices of exposition that are so ingenious and witty that they end by becoming tedious at times. "A mixture of the pedant and the *précieux* calculated to go down well with the bourgeoisie and the provinces," somebody has said.

He did not merely popularise the scientific achievement of the seventeenth century. It is important to note that the literary man intervenes at this crucial stage of the story and performs a second function—the translation of the scientific achievement into a new view of life and of the universe. Many of the scientists of the seventeenth century had been pious Protestants and Catholics, and in this very period both Robert Boyle and Isaac Newton showed considerable fervour concerning their Christianity

—even Descartes had thought that his work would serve the cause of religion. It had almost been a mystical urge and a religious preoccupation which had impelled a man like Kepler to reduce the universe to mechanical law in order to show that God was consistent and reasonable—that He had not left things at the mercy even of his own caprice. Fontenelle, as I have suggested, was something of a Talleyrand—charming, sceptical and evasive, with a little of both the pessimist and the cynic in his attitude to human nature. He had held his sceptical views before coming into touch with the scientific movement at all—had learned them from Lucretius and from more modern writers like Machiavelli and Montaigne. A scepticism which really had a literary genealogy combined to give to the results of the seventeenth-century scientific movement a bias which was rarely to be seen in the scientists themselves and which Descartes would have repudiated. And this was encouraged by the obstructive attitude of the Roman Catholic clergy in France, who helped to strengthen the impression that the Church was the enemy of scientific discovery and, indeed, of anything new. In this connection it is important to note that the great movement of the eighteenth century was a literary one—it was not the new discoveries of science in that epoch but, rather, the French *philosophe* movement that decided the next turn in the story and determined the course Western civilisation was to take. The discoveries of seventeenth-century science were translated into a new outlook and a new world-view, not by scientists themselves, but by the heirs and successors of Fontenelle.

So, in the great transition from the seventeenth century to the eighteenth century, there is not just a straight organic development of human thought—at a number of points there is, so to speak, a hiatus in the transmission, and these curious faults in the connection, these discontinuities, throw light on the whole structure and fabric of our general history. There is a break in the generations—the young reacting against the ideas and the educational system of their fathers. There is that appeal which Galileo was already making and which came to its climax in

Fontenelle and his successors—the appeal against the learned world of the time, against both Church and universities, to a new arbiter of human thought: a wider general reading-public. There is a further discontinuity when we find that it is the literary men who, sometimes in a rapid manner, perform the momentous task of translating the results of scientific work into a new general outlook, a new world-view. Finally, there is the greatest discontinuity of all—the emergence to importance of a new class.

After a long struggle the French kings had achieved, as fully as ever they were to achieve, that work which it was the function of the institution of monarchy to carry out in European history—namely, to weld provincial units into national ones, to reduce the power of local tyrants, and to impose upon the parochialism of half-developed peoples the larger idea of the State. In the 1680s Louis XIV was at the height of his power in France; the disastrous consequences of his policy had not yet undermined his work, and, since the *Fronde,* the might of a rebellious nobility had been broken—since 1660 there had been a remarkable return to stability in France after a long period of distress and anarchy which had often put the entire monarchy into peril. It has been noted that this return to stability would have been itself sufficient to produce a remarkable revival of economic activity, even if there had never been a Colbert to organise it, so that too much has perhaps been attributed to the work of this statesman and to the directing hand of government. And it is possible that students of history do not sufficiently stress the primary importance of a mere return to stability amongst the factors which help civilisation to make important steps forward and which were operative in this particular period. Anglo-French jealousy in the economic field is very noticeable in the latter part of the seventeenth century, and if one surveys the scene at this point in the story one has the feeling that the industrial revolution was going to take place in France rather than in England. France at this time was the most thickly populated country in Europe; the West India trade, in particular, had made a great development; and it has been computed that by the close of the seventeenth century 400,000 people were drawing their livelihood, directly or indirectly, from colonial commerce. Literary references at this time show that a wave of speculation swept over the country long before the famous upheaval associated with the name of John Law early in the eighteenth century; just as in William III's reign there already existed in England some of that speculative fever which Holland had witnessed at an earlier period still. Not only had the sting been taken out of the old nobility, but new classes were gaining the influential position, the intellectual leadership.

Saint-Simon, the famous memoir-writer of this period, belonged to the old *noblesse* and wrote with all the prejudices of that class. He calls our attention to a side of the reign of Louis XIV that we too often overlook when he denounces it as "a long reign of the vile bourgeoisie." During the sixteenth century it was the nobility who headed the Huguenot cause in the civil wars and in that period the bourgeoisie had had little part in literature. During much of the seventeenth century ecclesiastical leadership was so strong in both politics and intellectual life—so many Cardinals in high office in the state, for example—that historians speak of the Catholic Renaissance in the France of this period. Now, however, the very texture of society is palpably changing. Colbert himself was the son of a draper. In the reign of Louis XIV, Corneille, Racine, Molière, Boileau, La Bruyère, Pascal, etc.—a whole host of distinguished personages—came from this class. We have seen further signs of its growing influence and intellectual leadership in the case of the scientists themselves. All these modern developments in the reign of Louis XIV would have been more remarkable if the latter half of the reign had not been disastrous for the kind of state which Colbert had been promoting; and if there had not been an aristocratic reaction after that. Even so, the period 1660 to 1760 has been called "the golden age of the bourgeoisie" in France.

When men like Galileo and Descartes had decided to write their works in the vernacular rather than in Latin, avowedly for the purpose of appealing against the learned world to an intelligent reading-public, that stroke of policy had itself been of some significance, and something like it had occurred in Reformation Germany. There is involved in it one of those discontinuities in the history of civilisation which we have noticed in other aspects of the subject—for in the transfer of the intellectual leadership, in the reference to a new arbiter in the realm of thought, some things in the intellectual heritage of a civilisation are liable to be lost, as we can see in the parallel case of the Marxist appeal to still another arbiter—the proletariat—at the present day. In the case of France in the period with which we are dealing, there are particular reasons why the middle classes were able to make their intellectual leadership effective.

With a curious humility, or a rare sense of the true values in life, the French bourgeoisie—half-ashamed of their class, and spurning the usual bourgeois ideals—showed an anxiety to leave the bustle and conflict of economic life at the first opportunity, to be satisfied with a safe but modest income, to build a fine *hôtel* and to ape the nobility. Instead of reinvesting their money in industry and commerce and training up their children to carry on their work in the parental firm, they would buy lands or *rentes* or one of the safe little official posts, of which a village of two thousand inhabitants could supply a remarkable number. Alternatively, they would move into the professions of law and medicine—escaping from the business world, which they despised in their hearts, as soon as they achieved a modest competence. Colbert himself complained of the loss which was caused to the economic development of France by this pattern of the bourgeois conduct of life in that country. Travellers would compare France unfavourably with Holland, where the children tended naturally to be brought up to carry on and enlarge the business of their fathers. As one French historian has said: a "hæmorrhage of capital removed it from business as soon as it had been created."

So there were forces at work making the bourgeoisie a numerous body, but often preventing single members of the class from becoming extraordinarily rich or economically powerful. On the one hand, the French commercial horizon was the narrower for all this—the spirit of enterprise was limited and the safe "fixed income" became the homely ideal of the middle classes—you had the psychology of a nation of *rentiers.* And perhaps the reason for the extraordinary economic success of the Huguenots was due in part to the fact that they had less chance of becoming functionaries in the state or ornaments in society, so that they tended to stay in business. On the other hand, you can say that the French middle classes had a sense of true values—making enterprise subordinate to the purpose of "the good life"—and that, if every nation had done the same, a fortunate brake would have checked a disproportionate and one-sided progress in life and society. At any rate, it is important for the general history of culture that there was in France a bourgeoisie which knew leisure and looked for the delights of social life and desired to patronise art and learning. That class was one which was little impressed by either authority or tradition, and Fontenelle, as well as later writers of the *philosophe* movement, adopted the policy of making the intellectual work palatable and easy—unlike the older forms of academic or scholastic controversy. Whereas "reason" had once been a thing that required to be disciplined by a long and intensive training, the very meaning of the word began to change—now any man could say that he had it, especially if his mind was unspoiled by education and tradition. "Reason," in fact, came to signify much more what we today should call common sense.

The transition to the modern outlook and the birth of the *philosophe* movement does not proceed from the scientific revolution by what you might call normal ascent—by a direct logical development of ideas pure and simple. A number of things in our intellectual tradition were undoubtedly lost for the time being—one could write a whole history even of those things which have been recaptured into our tradition since that time, or those cases in which we have had to rediscover the meaning of ideas that during a considerable period had been dismissed as useless. In addition, the whole transition was achieved by intellectual conflict—which involved passions, misunderstanding and cross-purposes—and those who were fighting the obscurantism of universities, priests and provincially-minded aristocrats were tempted to be cavalier on occasion—they had no time to worry if there were a few unnecessary casualties in the course of the struggle. It is a curious fact, furthermore, that France gained the intellectual leadership in Europe through the prestige of an imposing collection of classical writers whom we associate with the court of Versailles. The movement which we have been studying takes place behind that imposing façade—almost in the shadows—and one historian of literature gathers it up under the title "The Decline of the Age of Louis XIV," though from the point of view which we are taking at the moment it was this movement which was the really germinal one. What is interesting here is another curious cultural discontinuity—the intellectual leadership which France had acquired as a result of her brilliance in one kind of literature she used in the eighteenth century to disseminate a different type of civilisation altogether.

There is one further field in which the intellectual changes of Louis XIV's reign touch the history of science—especially as they represent the extension of the scientific method into other realms of thought. In this case the field is that of politics, and all historians note its importance as the beginning of a development that led to the French Revolution. If on the one hand the French monarchy achieved its function in Louis XIV's reign, as completely as ever it was to do so, on the other hand we meet the beginnings of the criticism of the French monarchy—not the mere carping and obstructionism of the privileged classes now, but acute criticism from those sections of the French intelligentsia who could claim to understand the idea of the state better than the king himself. After reaching its peak in the 1680s, the reign of Louis XIV entered on that sinister decline by which it is principally remembered in this country; and from 1695 to 1707 a whole series of criticisms of a modern kind were levelled against the monarchy. The funeral orations of Fontenelle call attention to an aspect of this movement which is often overlooked—that is to say, the initial effect of the new scientific movement on political thought.

The political reformers in question were not yet the ideologues of later times—not doctrinaire writers after the manner of the *philosophes.* They wrote out of their genuine experience, and Fontenelle, who had to deal with various people of this type (since they were either members or honorary members of the *Académie des Sciences* in one capacity or another), called attention to the effect of the scientific movement upon them. The first result—the natural result—of the transfer of scientific methods to politics, as Fontenelle makes clear, was the insistence that politics requires the inductive method, the collection of information, the accumulation of concrete data and statistics. Fontenelle points out, for example, that it is necessary for the master of a state to study his country as a geographer or a scientist would. He describes with approval how Vauban, the great military engineer and one of the critics of Louis XIV, travelled over France, accumulating data, seeing the condition of things for himself, studying commerce and the possibilities of commerce, and gaining a knowledge of the variety of local conditions. Vauban, says Fontenelle, did more than anybody else to call mathematics out of the skies and attach it to various kinds of mundane utility. Elsewhere Fontenelle says with some exaggeration that it is practically to Vauban alone that modern statistics are due—Vauban, indeed, put statistics to the service of modern political economy and first applied the rational and experimental method in matters of finance. Similarly Fontenelle tells us that in England Sir William Petty, the author of *Political Arithmetic,* showed how much of the knowledge requisite for government reduces itself to mathematical calculation. Even where the movement was veering very much over to doctrinairism in the case of Fontenelle's friend, the Abbé de Saint-Pierre, we find an interesting proposal being put forward. Saint-Pierre desired to establish a body of scientific politicians to examine all kinds of projects for the improvement of the methods of government or the better conduct of economic affairs. Bureaux of experts would either conduct the various branches of government or would be attached to the various ministries in order to give inspiration and advice. Saint-Pierre appears to have been ready to encourage any member of the public to submit ameliorative projects for the consideration of the government. When the intellectual leadership passed in the eighteenth century to literary men, educated in classics and rhetoric, attention was turned away from this form of scientific politics, and political writing took a course to which in general we can hardly avoid giving the description "doctrinaire."

In an essay on "The Utility of Mathematics" Fontenelle stated a general doctrine which was coming to be widely held:

> The geometrical spirit is not so tied to geometry that it cannot be detached from it and transported to other branches of knowledge. A work of morals or politics or criticism, perhaps even of eloquence, would be better (other things being equal) if it were done in the style of a geometer. The order, clarity, precision and exactitude which have been apparent in good books for some time might well have their source in this geometric spirit. . . . Sometimes one great man gives the tone to a whole century; [Descartes], to whom might legitimately be accorded the glory of having established a new art of reasoning, was an excellent geometer.

If Descartes was the great fashion in the age of Louis XIV, Bacon was chosen as the patron saint, so to speak, of the French encyclopædists, and both of these authorities, as we have seen, encouraged the disposition to overhaul every kind of traditional teaching, and question the whole intellectual heritage. Fontenelle provides an example of the transfer of the scientific spirit, and the application of methodical doubt, in another work of his, on the *History of Oracles*. In a sense he is one of the precursors of the comparative method in the history of religion—the collection of myths of all lands to throw light on the development of human reason. In order to learn more about the primitive stages of our own history he recommends the study of primitive tribes as they exist in our own day—the Red Indians or the Lapps. He treats myths as, so to speak, a natural product, subject to scientific analysis—not the fruits of conscious imposture but the characteristic of a certain stage in human development. The human mind he regards as essentially the same in all times and ages, but subject to local influences—affected by the stage of social development so far achieved, the character of the country itself and the climate under which the human being happened to be living. He uses classical history or the stories of sailors and travellers in the port of Rouen, or Jesuit missionary narrations, as materials for a comparative study of myths. Here is a self-conscious attempt to show how the scientific method could receive extended application and could be transferred from the examination of purely material phenomena even into the field of what we should call human studies. And it was important that the methodical doubt, upon which Descartes had insisted at a very high level—and with peculiar implications as well as under a particularly strict discipline, as we have seen—was a thing easily vulgarised, a thing already changing its character in the age of Fontenelle, so that it had come to mean simply an ordinary unbelieving attitude, the very kind of scepticism which he had tried to guard against.

Chapter 10

The Place of the Scientific Revolution in the History of Western Civilisation

IT WAS THE PASSION of Ranke, whatever period or episode in history he might be studying, to seek to put it in its place in what he called "Universal History," which was the home you reached—the ocean you finally gazed upon —if you went far enough in your reflection upon a piece of narrative. To such a degree did he set his mind on this object that he could describe the quest for "the ocean of universal history" as the great purpose of his life, the ultimate goal of all his studies. It is strange that this—one of the most insistent parts of his message—should have been the one we have allowed to drop most completely out of sight in our usual studies of history, so that we tend to overlook it even when we are making an estimate of Ranke himself. Having examined many aspects of the seventeenth-century intellectual movement internally, however, it may be useful for us if we enlarge our perspective, standing some distance away from the story that we have been studying, and try to find the bearings of these events on the whole history of Western civilisation.

Until a comparatively recent date—that is to say, until the sixteenth or the seventeenth century—such civilisation as existed in our whole portion of the globe had been centred for thousands of years in the Mediterranean, and during the Christian era had been composed largely out of Græco-Roman and ancient Hebrew ingredients. Even at the Renaissance, Italy still held the intellectual leadership in Europe, and, even after this, Spanish culture had still to come to its climax, Spanish kings ruled over one of the great empires of history, and Spain had the ascendancy in the Counter-Reformation. Until a period not long before the Renaissance, the intellectual leadership of such civilisation as existed in this quarter of the globe had remained with the lands in the eastern half of the Mediterranean or in empires that stretched farther still into what we call the Middle East. While our Anglo-Saxon forefathers were semi-barbarian, Constantinople and Baghdad were fabulously wealthy cities, contemptuous of the backwardness of the Christian West.

In these circumstances it requires to be explained why the West should have come to hold the leadership in this part of the world; and, considering the Græco-Roman character of European culture in general, it is necessary to account for the division of the continent and to show why there should ever have arisen anything which we could call the civilisation of the West. Explanations are not difficult to find. Even when the Roman Empire surrounded the whole of the Mediterranean, there had been tension between East and West—a tension greatly increased when a second capital of the Empire had been founded and the oriental influences were able to gather themselves together and focus their influence on the city of Constantinople. In the subsequent period—the age of the Barbarian Invasions—the differences were increased when Constantinople held out against attacks and preserved the continuity of classical culture, while, as we have seen, the West was so reduced that it had to spend centuries recapturing and reappropriating it—gathering the fragments together again and incorporating them into its own peculiar view of life. The religious cleavage between Rome and Byzantium in the middle ages (when differences of religion seemed to penetrate into every department of thought) accentuated the discrepancies between Latin and Greek and led to divergent lines of development—in the West, for example, the friction between Church and State gave a tremendous stimulus to the progress of society and the rise of political thought. The West developed independently, then, but, though it may have been more dynamic, it was still for a long time backward. Even in the fifteenth century—in the period of the high Renaissance—the Italians were ready to sit at the feet of exiled teachers from Constantinople and to welcome them as men like Einstein were welcomed not very long ago in England or America. By this time, however, visitors from the Byzantine Empire were expressing their wonder at the technological advances in the West.

An important factor in the decline of the East and the rise of Western leadership, however, was one which has been unduly overlooked in our historical teaching, for it has played a decisive part in the shaping of the map of Europe, as well as in the story of European civilisation itself. From the fourth to the twentieth century one of the most remarkable aspects of the story—the most impressive conflict that spans fifteen hundred years—is the conflict between Europe and Asia, a conflict in which down to the time of Newton's *Principia* it was the Asiatics who were on the aggressive. From the fourth to the seventeenth century—when they still expected to reach the Rhine—the greatest menace to any culture at all in Europe were the hordes of successive invaders from the heart of Asia, coming generally by a route to the north of the Black Sea (a region which remained therefore a sort of no-man's-land almost down to the time of the French Revolution), but coming later south of the Caspian Sea and into Asia Minor and the Mediterranean region. Beginning with the Huns, and continuing with the Avars, the Bulgars, the Magyars, the Petchenegs, the Cumans, etc., these hordes—generally Turkish or Mongol in character—sometimes succeeded one another so quickly that one group was thrust forward into Europe by the pressure of others in the rear, or a chain of them would be jostling one another in a westerly direction—all of which culminated in the Mongol invasions of the thirteenth century, and the conquests of the Ottoman Turks after that.

These Asiatic invaders had something to do with the downfall of Rome and the western empire over fifteen hundred years ago; they overthrew Constantinople, the second Rome, in 1453; and for centuries they virtually enslaved Russia and dominated Moscow, which later came to stand in the position of a third Rome. It was they who hung as a constant shadow over the East and eventually turned the eastern Mediterranean lands into desert; and they put an end to the glory of Baghdad. Because of their activity over so many centuries it was the western half of Europe that emerged into modern history as the effective heir and legatee of the Græco-Roman civilisation. From the tenth century A.D. these Asiatics—though for centuries they had had tormented us and carried their depredations as far as the Atlantic coast—were never able to break into the West again or to do more than besiege Vienna. The tenth century represents something like a restoration of stability, therefore—the time from which Western civilisation makes its remarkable advance. One aspect of the period that comes to its culmination in the Renaissance is the emergence of western Europe to a position of independence and, indeed, of conscious cultural leadership.

A primary aspect of the Renaissance, however, as we have seen, is the fact that it completes and brings to its climax the long process by which the thought of antiquity was being recovered and assimilated in the middle ages. It even carries to what at times is a ludicrous extreme the spirit of an exaggerated subservience to antiquity, the spirit that helped to turn Latin into a dead language. Ideas may have appeared in new combinations, but we cannot say that essentially new ingredients were introduced into our civilisation at the Renaissance. We cannot say that here were intellectual changes calculated to transform the character and structure of our society or civilisation. Even the secularisation of thought which was locally achieved in certain circles at this time was not unprecedented and was a hot-house growth, soon to be overwhelmed by the fanaticism of the Reformation and the Counter-Reformation. During much of the seventeenth century itself we can hardly fail to be struck, for example, by the power of religion both in thought and in politics.

People have talked sometimes as though nothing very new happened in the seventeenth century either, since natural science itself came to the modern world as a legacy from ancient Greece. More than once in the course of our survey we ourselves have even been left with the impression that the scientific revolution could not take place—that significant developments were held up for considerable periods—until a further draft had been made upon the thought of antiquity and a certain minimum of Greek science had been recovered. Against all this, however, it might be said that the course of the seventeenth century, as we have studied it, represents one of the great episodes in human experience, which ought to be placed —along with the exile of the ancient Jews or the building-up of the universal empires of Alexander the Great and of ancient Rome—amongst the epic adventures that have helped to make the human race what it is. It represents one of those periods when new things are brought into the world and into history out of men's own creative activity, and their own wrestlings with truth. There does not seem to be any sign that the ancient world, before its heritage had been dispersed, was moving towards anything like the scientific revolution, or that the Byzantine Empire, in spite of the continuity of its classical tradition, would ever have taken hold of ancient thought and so remoulded it by a great transforming power. The scientific revolution we must regard, therefore, as a creative product of the West—depending on a complicated set of conditions which existed only in western Europe, depending partly also perhaps on a certain dynamic quality in the life and the history of this half of the continent. And not only was a new factor introduced into history at this time amongst other factors, but it proved to be so capable of growth, and so many-sided in its operations, that it consciously assumed a directing rôle from the very first, and, so to speak, began to take control of the other factors—just as Christianity in the middle ages had come to preside over everything else, percolating into every corner of life and thought. And when we speak of Western civilisation being carried to an oriental country like Japan in recent generations, we do not mean Græco-Roman philosophy and humanist ideals, we do not mean the Christianising of Japan, we mean the science, the modes of thought and all that apparatus of civilisation which were beginning to change the face of the West in the latter half of the seventeenth century.

Now I think it would be true to say that, for the historian, as distinct perhaps from the student of pre-history, there are not in any absolute sense civilisations that rise and fall—there is just the unbroken web of history, the unceasing march of generations which themselves overlap with one another and interpenetrate, so that even the history of science is part of a continuous story of mankind going back to peoples far behind the ancient Greeks themselves. But we cannot hold our history in our minds without any landmarks, or as an ocean without fixed points, and we may talk about this civilisation and that as though they were ultimate units, provided we are not superstitious in our use of the word and we take care not to become the slaves of our terminology. Similarly, though everything comes by antecedents and mediations—and these may always be traced farther and farther back without the mind ever coming to rest—still, we can speak of certain epochs of crucial transition, when the subterranean movements come above ground, and new things are palpably born, and the very face of the earth can be seen to be changing. On this view we may say that in regard not merely to the history of science but to civilisation and society as a whole the transformation becomes obvious, and the changes become congested, in the latter part of the seventeenth century. We may take the line that here, for practical purposes, our modern civilisation is coming out in a perceptible manner into the daylight.

In this period the changes were not by any means confined to France, though what we have hitherto studied has drawn our attention to certain aspects of the transition in the case of that country in particular. The movement was localised, however, and it is connected with the humming activity which was taking place, say from 1660, not only in England, Holland and France, but also actually between these countries—the shuttle running to and fro and weaving what was to become a different kind of Western culture. At this moment the leadership of civilisation may be said to have moved in a definitive manner from the Mediterranean, which had held it for thousands of years, to the regions farther north. There had been a pull in this direction on the part of the university of Paris in the later middle ages, and a still stronger pull after the Renaissance, when Germany had revolted against Rome and the north had taken its own path at the Reformation. In any case the Mediterranean had become at times almost a Mohammedan lake, and the geographical discoveries had been transferring the economic predominance to the Atlantic seaboard for a number of generations. For a moment, then, the history of civilisation was focused on the English Channel, where things were weaving themselves into new patterns, and henceforward the Mediterranean was to appear to the moderns as a backward region. Not only did England and Holland hold a leading position, but that part of France which was most active in promoting the new order was the Huguenot or ex-Huguenot section, especially the Huguenots in exile, the nomads, who played an important part in the intellectual exchange that was taking place. After 1685—after the Revocation of the Edict of Nantes—the alliance between the French and the English Protestants became more close. Huguenots fled to England or became the intermediaries for the publication in Holland of journals written in French and communicating English ideas. As the eighteenth century proceeded, the balance in Europe shifted more definitely to the north, with the rise of the non-Catholic powers of Russia and Prussia. Even in the new world it was the northern half of the continent that came to the forefront, and it was soon decided that this northern part should be British not French, Protestant not Roman Catholic—an ally, therefore, of the new form of civilisation. The centre of gravity of the globe itself seemed to have changed and new areas of its surface found for a time their "place in the sun."

This new chapter in the history of civilisation really opened when in 1660, after a long period of internal upheaval and civil war, a comparative political stability was brought about not merely in France but in general throughout the continent, where on all sides the institution of monarchy had been gravely challenged but had managed to reassert itself and to re-establish a public order. In fact, what we have already noticed in the case of France was still more true in England and Holland in the seventeenth century—we see the power in intellectual matters of what, in spite of the objections to the term, we must call the middle class. And just as the Renaissance was particularly associated with city-states (or virtual city-states) in Italy, South Germany and the Netherlands, where the commerce and economic development had produced an exhilarating civic life, so in the last quarter of the seventeenth century the intellectual changes were centred on the English Channel, where commerce had been making so remarkable a rise and so much prosperity seemed to have been achieved. The city-state disappeared from history in the first half of the sixteenth century; but on the wider platform of the nation-state the future still belonged to what we call the middle classes.

If we have in mind merely the intellectual changes of the period we are considering, they have been described by one historian under the title, *La crise de la conscience européenne*—a title which itself gives some indication of the importance of the transition that was taking place. What was in question was a colossal secularisation of thought in every possible realm of ideas at the same time, after the extraordinarily strong religious character of much of the thinking of the seventeenth century. John Locke produces a transposition into secular terms of what had been a presbyterian tradition in political thought, and in doing so he is not a freak or a lonely prophet—he stands at the pivotal point in what is now a general transition. This secularisation came at the appropriate moment for combination with the work of the scientific revolution at the close of the seventeenth century; yet it would appear that it was not itself entirely the result of the scientific achievements—a certain decline of Christianity appears to have been taking place for independent reasons. One is tempted to say on quite separate grounds that this period emerges as one of the lowest points in the history of Western Christianity between the eleventh century and the twentieth. If we look at the general moral tone of Charles II's reign after the period of the Puritan ascendancy and compare it with the extraordinarily parallel case of the Regency in France after the religiosity of the closing years of Louis XIV's reign, it is difficult to resist the feeling that in both cases a general relaxation in religion and morals followed periods of too great tension—these things were not the straight results of the scientific revolution taken in isolation. In any case it lay perhaps in the dialectic of history itself that in the long conflicts between Protestant and Catholic the secular state should rise to independence and should secure an arbitral position over what now seemed to be mere religious parties within it. The whole story of the Renaissance shows within the limits of the city-state how the exhilarating rise of an urban civilisation is liable to issue in a process of secularisation—the priest as well as the noble loses the power that he was able to possess in a more conservative agrarian world. Something parallel has happened over and over again in the case of nation-states when not only have towns become really urban in character—which is late in the case of England, for example—but when a sort of leadership in society has passed to the towns, and literature itself comes to have a different character.

There is another reason why it would be wrong to impute all the changes in thought at this time to the effect of the scientific discoveries alone. It happened that just at this moment books of travel were beginning to have a remarkable effect on the general outlook of men—a postponed result of the geographical discoveries and of the growing acquaintance with distant lands. Western Europe was now coming to be familiar with the widespread existence of peoples who had never heard of ancient Greece or of Christianity. When these were taken into one's larger survey, the European outlook came to be envisaged not as universal, not necessarily even as central, but somewhat as a regional affair. It became possible to look upon it as only the local tradition of a comparatively small section of the globe. So one could begin to regard one's own culture, even one's own religion, with a great degree of relativity. It was possible to look on each local creed as embodying one essential truth, but covering that truth with its own local myths, perversions and accretions. What was common to all was the universal irreducible truth—the principles of natural religion—and in French books of travel, therefore, you find the essential ingredients of Deism before John Locke had shown the way. Furthermore, you could feel that in Western Europe Christianity had its basis in the same universal truth, but the principles had been covered (in Roman Catholicism, for example) by local accretions, revelations and miracles, from which it now required to be extricated. The results of all this harmonised with the operations of the new science, and strengthened the case for the kind of Deism which Newton's system seemed to encourage—a Deism which required a God only at the beginning of time to set the universe in motion.

From this period also there developed in a remarkable way and with extraordinary speed the tendency to a new type of Protestantism—the more liberal type which most of us have in mind when we are in controversy on this subject. It was a Protestantism married to the rationalising movement, and so different from the original Protestantism that it now requires an effort of historical imagination to discover what Martin Luther had in mind. Some remarkable developments in this rationalising tendency were only checked in England by the rise and the pervasive influence of John Wesley, who, however, also carries so many of the features of the Age of Reason in himself. On the other hand we have to note that if books of travel affected the attitude of western Europeans to their own traditions, the very attitude these people adopted (the kind of relativity they achieved) owed something to a certain scientific outlook which was now clearly becoming a more general habit of mind. Similarly, when in the 1660s a writer like Joseph Glanvill could produce a book on *The Vanity of Dogmatising,* insisting on the importance of scepticism in science and on the system of methodical doubt, it is impossible to deny that this critical outlook is an effect of the scientific movement. In general, we ought not to close our eyes to the extremely dislocating effects of that general overthrow of the authority of both the middle ages and antiquity which again had been produced by the scientific revolution. Either we may say, therefore, that a number of converging factors were moving the Western world in one prevailing direction, or we must say that there was one wind so overpowering that it could carry along with it anything else that happened—a wind so mighty that it gathered every other movement into its sweep, to strengthen the current in favour of secularisation at this time.

The changes which took place in the history of thought in this period, however, are not more remarkable than the changes in life and society. It has long been our tendency to push back the origins of both the industrial revolution and the so-called agrarian revolution of the eighteenth century, and though, as I have said, we can trace back the origin of anything as far as we like, it is towards the end of the seventeenth century that the changes are becoming palpable. The passion to extend the scientific method to every branch of thought was at least equalled by the passion to make science serve the cause of industry and agriculture, and it was accompanied by a sort of technological fervour. Francis Bacon had always laid stress on the immense utilitarian possibilities of science, the advantages beyond all dreams that would come from the control of nature; and it is difficult, even in the early history of the Royal Society, to separate the interest shown in the cause of pure scientific truth from the curiosity in respect of useful inventions on the one part, or the inclination to dabble in fables and freakishness on the other. It has become a debatable question how far the direction of scientific interest was itself affected by technical needs or preoccupations in regard to shipbuilding and other industries; but the Royal Society followed Galileo in concerning itself, for example, with the important question of the mode of discovering longitude at sea. Those who wish to trace the development of the steam-engine will find that it is a story which really begins to be vivid and lively in this period. Apart from such developments, the possibilities of scientific experiment were likely themselves to be limited until certain forms of production and technique had been elaborated in society generally. Indeed, the scientific, the industrial and the agrarian revolutions form such a system of complex and interrelated changes, that in the lack of a microscopic examination we have to heap them all together as aspects of a general movement, which by the last quarter of the seventeenth century was palpably altering the face of the earth. The hazard consists not in putting all these things together and rolling them into one great bundle of complex change, but in thinking that we know how to disentangle them—what we see is the total intricate network of changes, and it is difficult to say that any one of these was the simple result of the scientific revolution itself.

Embraced in the same general movement is that growth of overseas trade which we have already noticed in the case of France—and once again we find a remarkable postponed result of the geographical discoveries of a much earlier period, reminding us that the New World represents one of the permanent changes in the conditioning circumstances of the modern age, one of the great standing differences between medieval and modern times, its results coming in relays and reproducing themselves at postponed periods. In the England of Charles II's reign we begin to see that we are an empire; the Board of Trade and Plantations comes to occupy a central position in the government; it is after 1660 that the East India Company reaps its colossal harvests. We begin to hear much less in the way of complaint about the excessive numbers of the clergy—henceforward what we begin to hear are complaints about the growing number of customs officials, Treasury men, colonial officers, contractors—all of them subject to corruption by the government. This is the epoch in which, as historians have long pointed out, wars of trade—especially amongst the Dutch, the French and the English—succeeded the long series of wars of religion. In a similar way we must take note of the foundation of such things as the Bank of England and the national debt—a new world of finance that alters not merely the government but the very fabric of the body politic. We have seen how in France and England there already existed signs of that speculative fever which culminated in the scheme of John Law on the one hand and the South

Sea Bubble on the other; while in Holland there had been a parallel financial sensation earlier still.

For two thousand years the general appearance of the world and the activities of men had varied astonishingly little—the sky-line for ever the same—so much so that men were not conscious of either progress or process in history, save as one city or state might rise by effort or good fortune while another fell. Their view of history had been essentially static because the world had been static so far as they could see—life in successive generations played out by human beings on a stage that remained essentially the same. Now, however, change became so quick as to be perceptible with the naked eye, and the face of the earth and the activities of men were to alter more in a century than they had previously done in a thousand years. We shall see later, in connection with the idea of progress, how in general—and for effective purposes—it was in this period that men's whole notion of the process of things in time was thrown into the melting-pot. And the publication of a host of journals in France, England and Holland speeded up the pace of intellectual change itself.

A curious feature of seventeenth-century English life illustrates the growing modernity of the world, and throws light not only on social change but on a certain different flavour that is becoming apparent in the prevailing mentality. There is a foretaste of it in the debates of James I's reign when we find that certain people called Projectors are being attacked in parliament—the sort of people whom we might call company-promoters, and who devised schemes for making money. They developed very greatly after the Restoration, becoming a considerable phenomenon in William III's reign, and they culminated in the period of the South Sea Bubble, when companies were founded to execute all kinds of fantastic schemes, including a method of procuring perpetual motion. Just before the end of the seventeenth century Daniel Defoe—who emerges as a remarkably modern mind—produced an *Essay on Projects* in which he commented on the whole phenomenon, satirised the Projectors, but then produced many schemes of his own to swell the flood. It is curious to note that these Projectors provided another of what we should call the "mediations" which assisted the passage to the *philosophe* movement; for though some of them had schemes for getting rich quickly—Defoe had a scheme for improving trade by settling the problem of the Barbary pirates, for example—some others had wider views: schemes of general amelioration, schemes for tackling the problem of the poor, plans for female education, devices for getting rid of the national debt. The famous socialistic system of Robert Owen was taken, as Owen himself explains, from John Bellairs, who produced the design of it in 1696 under the title of "a scheme by which the rich were to remain rich and the poor were to become independent, and children were to be educated." Bellairs had other proposals for general amelioration—for example in connection with prison-reform. Such things easily passed into projects for new forms of government, and curious mechanical schemes were put forward—the prelude to modern constitution-making and blue-prints for Utopia. They make it clear that the historical process is very complex; that while the scientific movement was taking place, other changes were occurring in society—other factors were ready to combine with it to create what we call the modern world.

It is always easy for a later generation to think that its predecessor was foolish, and it may seem shocking to state that even after the first World War good historians could write the history of the nineteenth century with hardly a hint of the importance of Socialism, hardly a mention of Karl Marx—a fact which we should misinterpret unless we took it as a reminder of the kind of faults to which all of us are prone. Because we have a fuller knowledge of after-events, we today can see the nineteenth century differently; and it is not we who are under an optical illusion—reading the twentieth century back unfairly into the nineteenth—when we say that the student of the last hundred years is missing a decisive factor if he overlooks the rise of Socialism. A man of insight could have recognised the importance of the phenomenon long before the end of the nineteenth century. But we, who have seen the implications worked out in the events of our time, need no insight to recognise the importance of this whole aspect of the story.

Something similar to this is true when we of the year 1957 take our perspective of the scientific revolution—we are in a position to see its implications at the present day much more clearly than the men who flourished fifty or even twenty years before us. And, once again, it is not we who are under an optical illusion—reading the present back into the past—for the things that have been revealed in the 1950s merely bring out more vividly the vast importance of the turn which the world took three hundred years ago, in the days of the scientific revolution. We can see why our predecessors were less conscious of the significance of the seventeenth century—why they talked so much more of the Renaissance or the eighteenth-century Enlightenment, for example—because in this as in so many other cases we can now discern those surprising overlaps and time-lags which so often disguise the direction things are taking. Our Græco-Roman roots and our Christian heritage were so profound—so central to all our thinking—that it has required centuries of pulls and pressures, and almost a conflict of civilisations in our very midst, to make it clear that the centre had long ago shifted. At one time the effects of the scientific revolution, and the changes contemporary with it, would be masked by the persistence of our classical traditions and education, which still decided so much of the character of the eighteenth century in England and in France, for example. At another time these effects would be concealed through that popular attachment to religion which so helped to form the character of even the nineteenth century in this country. The very strength of our conviction that ours was a Græco-Roman civilisation—the very way in which we allowed the art-historians and the philologists to make us think that this thing which we call "the modern world" was the product of the Renaissance—the inelasticity of our historical concepts, in fact—helped to conceal the radical nature of the changes that had taken place and the colossal possibilities that lay in the seeds sown by the seventeenth century. The seventeenth century, indeed, did not merely bring a new factor into history, in the way we often assume—one that must just be added, so to speak, to the other permanent factors. The new factor immediately began to elbow the other ones away, pushing them from their central position. Indeed, it began immediately to seek control of the rest, as the apostles of the new movement had declared their intention of doing from the very start. The result was the emergence of a kind of Western civilisation which when transmitted to Japan operates on tradition there as it operates on tradition here—dissolving it and having eyes for nothing save a future of brave new worlds. It was a civilisation that could cut itself away from the Græco-Roman heritage in general, away from Christianity itself—only too confident in its power to exist independent of anything of the kind. We know now that what was emerging towards the end of the seventeenth century was a civilisation exhilaratingly new perhaps, but strange as Nineveh and Babylon. That is why, since the rise of Christianity, there is no landmark in history that is worthy to be compared with this.

Chapter 11

The Postponed Scientific Revolution in Chemistry

IT HAS OFTEN BEEN a matter of surprise that the emergence of modern chemistry should come at so late a stage in the story of scientific progress; and there has been considerable controversy amongst historians concerning the reasons for this. Laboratories and distilleries, the dissolution or the combination of substances and the study of the action of acid and fire—these things had been familiar in the world for a long time. By the sixteenth century there had been remarkable advances on anything that had been achieved in the ancient world in the field of what might be called chemical technology—the smelting and refining of metals, the production and the treatment of glass-ware, pottery and dyes, the development of such things as explosives, artists' materials and medicinal substances. It would appear that experimentation and even technological progress are insufficient by themselves to provide the basis for the establishment of what we should call a "modern science." Their results need to be related to an adequate intellectual framework which on the one hand embraces the observed data and on the other hand helps to decide at any moment the direction of the next enquiry. Alchemy had certainly failed to produce the required structure of scientific thought, and perhaps even in the experimental field it was a borrower rather than a contributor, playing a less important part, therefore, in the evolution of chemistry than was once imagined. From the early sixteenth century, the more genuine precursors were the "iatro-chemists," who followed Paracelsus in their insistence on the importance of chemical remedies for the physician. And, until late in the eighteenth century, chemistry came to be particularly associated with the practice and the teaching of medicine.

Robert Boyle had set out to bring about a marriage between the chemical practitioner and the natural philosopher; and from this time the story does at least become more comprehensible to us—there are recognisable aspirations in the directions of science, with less of what to us seems mere capriciousness or mystification. Boyle's fame was great; the Latin editions of his works were numerous; some aspects of his researches certainly had influence upon the Continent. Englishmen in his time were beginning to be particularly drawn to the type of problem that was to be important throughout the following century. We have already seen, however, that Boyle's fervour for the "mechanical philosophy" may have had an unfortunate effect upon his work at what he regarded as the crucial point. At the same time his Baconian method—his love of describing experiments independent of explanation or synthesis—may also have worked, though in a contrary direction, to put a limit upon his influence. Joseph Freind, Professor of Chemistry in the University of Oxford, wrote in 1712:

> Chemistry has made a very laudable progress in Experiments; but we may justly complain, that little Advances have been made towards the explication of 'em. . . . No body has brought more Light into this Art than Mr. Boyle . . . who nevertheless has not so much laid a new Foundation of Chemistry as he has thrown down the old.

When we study the history of science, it is useful to direct our attention to the intellectual obstruction which, at a given moment, is checking the progress of thought—the hurdle which it was then particularly necessary for the mind to surmount. In mechanics, at the crucial moment, as we have seen, it had been the very concept of motion; in astronomy, the rotation of the earth; and in physiology, the movement of the blood and the corresponding action of the heart. In chemistry, once again, it would seem that the difficulty in this period lay in certain primary things which are homely and familiar—things which would not trouble a schoolboy in the twentieth century, so that it is not easy for us to see why our predecessors should seem to have been so obtuse. It was necessary in the first place that they should be able to identify the chemical elements, but the simplest examples were perhaps the most difficult of all. For thousands of years, air, water and fire had been wrapped up in a myth somewhat similar to the myth of the special ethereal substance out of which the heavenly bodies and celestial spheres were thought to have been made. Of all the things in the world, air and water seemed most certain to be irreducible elements, if indeed—as Van Helmont suggested—everything in the world could not be resolved into water. Even fire seemed to be another element—hidden in many substances, but released during combustion, and visibly making its escape in the form of flame. Bacon and some of his successors in the seventeenth century had conjectured that heat might be a form of motion in microscopic particles of matter. Mixed up with such conjectures, however, we find the view that it was itself a material substance; and this latter view was to prevail in the eighteenth century. Men who had made great advances in metallurgy and had accumulated much knowledge of elaborate and complicated chemical interactions, were as yet unable to straighten out their ideas on these apparently simple topics. It would appear to us today that chemistry could not be established on a proper footing until a satisfactory starting-point could be discovered for the understanding of air and water; and for this to be achieved it would seem to have been necessary to have a more adequate idea both about the existence of "gases" and about the process of combustion. The whole development depended on the recognition and the weighing of gases; but at the opening of the eighteenth century there was no realisation of the distinctions between gases, no instrument for collecting a gas, and no sufficient consciousness of the fact that measurements of weight might play the decisive part amongst the data of chemistry.

From the time of Boyle and Hooke a great deal of

activity was being concentrated on the crucial and interrelated processes of combustion, calcination and respiration. A considerable amount of study had been devoted also to the air; and these two branches of enquiry had obvious relations with one another. Earlier in the century, Van Helmont had examined what in those days were regarded as "fumes," but though he discovered and described certain things which we should call "gases," he had regarded these as impurities and exhalations—as earthy matter carried by the air—and for him there was really only one "gas," which itself was only a form assumed by water, water being the basis of all material things. The contemporaries of Boyle had come near to discovering various gases, and were able to detect something which clearly suggests oxygen, while they talked of nitro-aerial particles with which they associated not only gunpowder but earthquakes and lightning, and even freezing, so that here was something which appeared to have almost a cosmic significance. They did not realise the existence of different gases, however, or understand that the air might comprise different gases; and it would be anachronistic to see them as the discoverers of oxygen and nitrogen. The problem of the air was to be elucidated only by a more methodical handling and a more acute examination of the processes of combustion. In this connection the emergence of the phlogiston theory provides a significant moment in the history of chemistry.

This theory, which was to become so fashionable for a time in the eighteenth century, embodied the essential feature of a tradition that went back to the ancient world—namely, the assumption that, when anything burns, something of its substance streams out of it, struggling to escape in the flutter of a flame, and producing a decomposition—the original body being reduced to more elementary ingredients. The entire view was based upon one of those fundamental conclusions of commonsense observation which (like Aristotle's view of motion) may set the whole of men's thinking on the wrong track and block scientific progress for thousands of years. The theory might have represented an advance at the time when it was first put forward; but in future ages no rectification seems to have been possible save by the process of going back to the beginning again. Under the system of the Aristotelians it was the "element" of fire which had been supposed to be released during the combustion of a body. During most of the seventeenth century it was thought to be a sulphurous "element"—not exactly sulphur as we know it, but an idealised or a mystical form of it—materially a different kind of sulphur in the case of the different bodies in which it might appear. A German chemist, J. J. Becher, who was contemporary with Boyle, said in 1669, that it was *terra pinguis*—an oily kind of earth; and at the opening of the eighteenth century another German chemist, G. E. Stahl, took over this view, elaborating it down to 1731, renaming the *terra pinguis* "phlogiston," and regarding phlogiston as an actual physical substance—solid and fatty, though apparently impossible to secure in isolation. It was given off by bodies in the process of combustion, or by metals in the process of calcination, and it went out in flame to combine with air, or perhaps deposited at least a part of itself in an unusually pure form as soot. If you heated the calx—the residue of a calcinated metal—along with charcoal, the substance would recover its lost phlogiston and would be restored to its original form as a metal. Charcoal was therefore regarded as containing much phlogiston, while a substance like copper was supposed to contain very little. This phlogiston theory was not everywhere immediately accepted: a man like the famous Boerhaave seemed able to ignore it; and some people who worked within the framework of it may have been little affected by it. The French were apparently going over to it in a general way from about 1730; but the rapid spread and development would seem to have occurred only in the 1740s and 1750s—it would appear to have been in about the middle of the century that the doctrine established itself as the orthodox one amongst chemists. The case has been made out that only some two decades later did it begin to occupy much place in the chemical literature; and that it caused the most stir in the world at the time when it was being seriously challenged.

It had been realised all the time, and it was known to Stahl, who really developed the phlogiston doctrine, that when burning had taken place or metals had been calcined an actual increase in weight had been discovered in the residue. The fact may have been known to the Arabs; it was realised by some people in the sixteenth century; it was brought to the attention of the Royal Society in London after 1660. In the seventeenth century the view had even been put forward more than once that, in the act of burning, a substance took something out of the air, and that this process of combination accounted for the increase which was observed in the weight. The phlogiston theory—the theory that something was lost to a body in the process of burning—is a remarkable evidence of the fact that at this time the results of weighing and measuring were not the decisive factors in the formation of chemical doctrine. Like Aristotle's view of motion, therefore, the phlogiston theory answered to certain *prima facie* appearances, but stood almost as an inversion of the real truth—a case of picking up the wrong end of the stick. It is remarkable how far people may be carried in the study of a science, even when an hypothesis turns everything upside-down, but there comes a point (as on the occasion when Aristotle reaches the problem of projectiles) where one cannot escape an anomaly, and the theory has to be tucked and folded, pushed and pinched, in order to make it conform with the observed facts. This happened in the case of the phlogiston theory when the scientist found it impossible to evade the fact of the augmented weight of bodies after combustion or calcination.

Somebody suggested that the phlogiston might have negative weight, a positive virtue of "levity," so that a body actually became heavier after losing it. Such an hypothesis, however, made a serious inroad on the whole doctrine of a solid phlogiston; and we can see that this ancient idea of "levity" had ceased to be capable of carrying much conviction by the eighteenth century. One

German chemist, Pott, suggested that the departure of the phlogiston increased the density of the substance which had held it, and J. Ellicott in 1780 put forward the view that its presence in a body "weakened the repulsion between the particles and ether," thereby diminishing their mutual gravitation." The more popular view seems to have been that while the burning produced a loss of phlogiston and a loss of weight, a secondary and somewhat incidental operation occurred, which more than cancelled the loss of weight. It is curious to find that Boyle had considerable influence at any rate in one of his errors, because he had noted the increase of weight when substances were burned and he had explained it by the suggestion of fire-particles which insinuated themselves into the minute pores of the burned matter, and which he regarded as having weight but as being able to pass through the glass walls of a closed container. Not only was this view held by some people in the eighteenth century, but it was possible to hold the phlogiston theory and still believe that weight was gained in combustion as a result of something which was taken incidentally from the air—this on a sufficient scale to override any reduction that had been produced by the loss of the phlogiston. For a considerable part of the eighteenth century the anomalies in phlogiston chemistry may be taken as an illustration of the insufficient attention given to the question of weight in the formation of hypotheses. Without the device of a secondary increase in weight, however, the phlogiston theory could not have put up the fight which it did in the closing decades of the century.

The phlogiston theory had a further disadvantage in that it carried the implication that nothing which could be burned or calcined could possibly be an element. Combustion implied decomposition. Only after the removal of the phlogiston could you expect to find matter in its elementary forms. If in calcination we today see oxygen combining with a metal, the eighteenth century saw the compound body—the metal—being decomposed and deprived of its phlogiston. If in the reverse process we see the oxygen being removed from a lead oxide to recover the original element, they imagined that they were adding something —restoring the phlogiston—so that the lead which emerged was a chemical compound, a product of synthesis. For men who worked on a system of ideas like this, it was not going to be easy to solve the problem of the nature of chemical elements.

Modern historical writers have tended to try to be kind to the phlogiston theory, apparently on the view that it is the historian's function to be charitable, and that the sympathy due to human beings can properly be extended to inanimate things. It has been noted that the men who established the theory made a mistake that was common in ancient times—they realised the existence of certain properties and turned these into an actual substance. One writer has said that the phlogiston theory "was the first important generalisation in chemistry correlating in a simple and comprehensive manner a great number of chemical actions and certain relations existing between a great variety of substances." As the unifying factor and the ground of the relations was the fictitious phlogiston, however, it is difficult to see how anything was facilitated. It is claimed that not only was there a phlogiston theory, but this theory of combustion gradually extended itself into a system of chemistry—what you have now is a period of phlogistic chemistry. And it is true that from 1750 we possess something more like a history of chemistry, whereas before we seem to have rather a history of chemists—too many of them standing on an independent footing with their separate theories—so that the general acceptance of phlogiston seemed to bring them all into one intellectual system. Some writers have pointed out that within the framework of phlogistic chemistry many experiments were carried out, and it would hardly have occurred to the enquirer to make some of these if a different framework had been currently accepted. Many important experiments had been carried out under preceding intellectual systems, however; and it might be held (though such speculations certainly have their dangers) that the emergence of chemistry as a science is remarkably late, that the chemistry of Boyle and Hooke may not have taken the shortest possible route to arrive at Lavoisier, and that the interposition of the phlogistic theory made the transition more difficult rather than more easy. This theory was in a significant sense a conservative one, though it may have had the effect of making the conservative view more manageable for a certain period.

It was made to serve comprehensive purposes. Because bodies changed colour at different degrees of heating it could be extended into an explanation of colour. But that in spite of this it created difficulties during the generation of its unquestioned predominance is shown by Lavoisier's taunts in the 1780s, when he said that phlogiston now had to be free fire and now had to be fire combined with an earthy element; sometimes passed through the pores of vessels and sometimes was unable to do so; and was used to explain at the same time causticity and non-causticity, transparency and opacity, colour and the absence of colour. Furthermore, the last two decades of the eighteenth century give one of the most spectacular proofs in history of the fact that able men who had the truth under their very noses, and possessed all the ingredients for the solution of the problem—the very men who had actually made the strategic discoveries—were incapacited by the phlogiston theory from realising the implications of their own work. Although it is true in the history of thought that false ideas or half-truths sometimes act as a convoying agency—leading the enquirer to a sounder form of generalisation, and then dropping out of the story when they have fulfilled their purpose—it still is not clear which of the famous discoveries of Black, Cavendish, Priestley and Lavoisier would have proved more difficult to achieve if the phlogiston theory had not existed. Perhaps it is possible to say that the perpetual removal of phlogiston from one body to another or from a body into the air itself, and then the return of the phlogiston to the original body, accustomed the mind of the chemist to the practice of moving and reshuffling the discs so that he became more agile—more ready to see elements ejected or transferred in the course of chemical reactions. But if chemistry made great advances from 1750 it is much more clear that the development of methods of collecting gases, and the demonstration by Joseph Black of what could be achieved by the use of the balance, together with the general improvement in the manufacture of apparatus (which was a serious and often an expensive matter at this time), are much more tangible and concrete causes of progress.

Although there appears to have been continued interest in chemistry and chemical experiments during the first half of the eighteenth century, it is perhaps true to say that no remarkable genius emerged to develop what had been achieved in the previous decades by Boyle, Hooke and Mayou. In Germany and Holland, where there was considerable interest in the application of science to the industrial arts, there was an awakening in the second quarter of the century, and the pupils of Boerhaave at Leyden carried his influence to the universities of many countries, one of them William Cullen, being the teacher of Joseph Black. In Britain around the middle of the century chemists were engaged very much with pharmacology or technology, or in pursuits that we should associate rather with physics. It has been pointed out that the Industrial Revolution in England depended "as much upon chemical discovery as upon mechanical discovery," with sulphuric acid playing a peculiarly important part in the story. Scotland appears to have been the seat of important developments on this side, Black performing in Edinburgh something like the rôle of Boerhaave in Leyden.

When the Swedish chemist, Scheele, embarked upon the problem of combustion he found that it would be impossible to arrive at satisfactory answers to his questions until he had dealt with the problem of the air, to which he devoted his attention in the years 1768-73. The fact that the two problems were related and that combustion had even a curious correspondence with respiration had long been realised, and there are hints of it in the ancient world. Certain chemists—certain Englishmen, for example—have already been mentioned, who in the seventeenth century had put forward suggestions on this question which were in advance of the views of the phlogistonists. In the seventeenth century, however, the problem had been made more difficult by ideas concerning the purely mechanical operation of the air, or concerning the action of the atmosphere as the mere receptacle for the fumes that were given out on combustion. It was held that if a lighted candle soon went out when enclosed under a container, the increasing pressure of the air loaded with fumes was responsible for extinguishing it. And even after the air-pump had been invented and it could be shown that the candle would not burn in a vacuum, a purely mechanistic theory was still possible—you could argue that the pressure of the air was necessary to force out the fire and flame from the burning substance, so that any rarifying of the atmosphere would rob the flame of its vital impulse. At the time of the phlogiston theory mechanistic ideas still prevailed, for it was eminently the function of the air to absorb the escaping phlogiston, and in time the air became saturated, which accounted for the extinguishing of the candle under a closed container.

Though the waters are muddy, and things apparently inconsistent with one another could co-exist, it would seem to be true that in the period before 1750 chemists did not think of the actual air as being a mixture, though they were well aware that the atmosphere might be loaded—more in some places than in others—with more or less obnoxious alien effluvia. Up to this date also they had no clear idea even of the possible existence of perfectly distinct kinds of gases. The differences which they observed on occasion they were liable to ascribe to a modification of what was fundamentally the same substance. In any case the atmosphere was so intangible and subtle that they found it difficult to imagine that air (or any part of it) might be "fixed" (as they called it)—trapped into combining with a solid substance to form a stable compound. They appear to have been more prepared to believe that particles of air might lurk as foreign bodies in the minute pores of solid substances, and that this might account for any increase in weight taking place after combustion.

Stephen Hales showed in 1727 that gases could be "fixed", however, and that in animal and vegetable life this process was constantly taking place. He discovered a way of collecting gases over water and would examine the quantity produced by chemical action from a given weight of materials. He even showed that the gases, or "airs" as he called them, which he collected from various substances, differed in regard to colour, smell, solubility in water, inflammability, etc., though no great importance was attached to these differences at the time. Hales, like his readers, still thought that what was in question was one single air under different conditions—"infected" or "tainted", as he said, with extraneous fumes or vapours. It was an important moment, therefore, when Joseph Black, in 1754, demonstrated the existence of an "air" which, unlike "common air," was attracted to quicklime, and which he studied in various combinations, though he did not isolate and collect it or provide a full account of its characteristics. He called it "fixed air"—the term which Hales had used—and he showed that not only could it exist in a free state but it could be captured into solid bodies; indeed, it could combine with one substance and then be transferred into combination with another. He noted soon afterwards that it differed from the "air" produced by the solution of metals in acids, and resembled common air that had been tainted by breathing or combustion. The method by which Black made this examination of what we should call carbon dioxide was as important as the discovery itself. His work stood as a model for the thorough and intensive study of a chemical reaction, and revealed the decisive results that could be achieved by the use of the balance. He showed that common air can be an actual participant in chemical processes and that an air could exist which was different from ordinary air. At the same time it does not appear that even he was fully conscious of the independent existence of separate gases. He seems to have been prepared to regard his "fixed air" as a modification of common air produced by the operation of the inflammable principle—namely, phlogiston.

In 1766 Henry Cavendish carried the story further in some studies of what he described in the words of Boyle as "factitious airs," which, he said, meant "any kind of air which is contained in other bodies in an inelastic sense and is produced from thence by art." Amongst other things he dissolved marble in hydrochloric acid, producing Black's "fixed air"; dried the gas and used the device of storing it over mercury, since it was soluble in water; and expanded the description of it, calculating its specific gravity, its solubility in water, etc. He also produced hydrogen by dissolving either zinc or iron or tin in sulphuric or hydrochloric acid, and found that there was no difference in the gas if he used different acids on the metals; and again he calculated its specific gravity. It was clear, therefore, that these two gases had a stable existence, and could be produced with permanent properties—they were not the capricious result of some more inconstant impurities in the air. And though both these gases had actually been discovered much earlier, they had not been separated in the mind from other things of a kindred nature—hydrogen had not been distinguished from other inflammable gases, for example. Even now, however, there was a feeling that in the last resort only one kind of air really existed—common air—and that the varieties were due to the presence or absence of phlogiston. Cavendish was inclined to identify his "inflammable air" with phlogiston, though there were objections to this, since phlogiston had been assumed to be not the burning body itself but a substance that left the burning body—and if the hydrogen was phlogiston how could phlogiston leave itself?

Joseph Priestley further improved the apparatus for collecting gases, and it is possible that, coming as an amateur without the means for any great outlay, he was driven to greater ingenuity in the devising of the requisite instruments. He had actually produced oxygen without realising it by 1771, and long before his time there had been ideas of a specially pure portion, or a specially pure constituent, of the air, which had been recognised as important for breathing and combustion. In 1774 Priestley isolated oxygen, but at first thought it to be what he called "modified" or "phlogisticated nitrous air," and what we call nitrous oxide. Later, for a moment, after further tests, he decided that it must be common air, but by the middle of March 1775 he realised that it was five or six times more effective than the ordinary atmosphere, and he named it "dephlogisticated air." The discovery had been made a few years earlier by the Swedish chemist, Scheele, who published his results later, but who showed more insight than Priestley by the way in which he recognised the existence of two separate gases in the air. Whoever may deserve the credit, the discovery and isolation of oxygen marks an important date in the history of chemistry.

By this time the position was coming to be complicated and chaotic. You have to remember that a deep prejudice regarded the air as a simple primordial substance, and a deeper prejudice still regarded water as an irreducible element. The balance of opinion, on the other hand, was in favour of regarding the metals as compounds, and if one of these, under the action of an acid, produced hydrogen, it was natural to think that the hydrogen had simply been released from the metal itself. When it was found later that an exploded mixture of hydrogen and oxygen formed water, it was simplest to argue that water was one of the constituents of oxygen, or of both the gases, and had been precipitated in the course of the experiment. When a gas was produced after the combustion of a solid body they gradually sorted out the fact that sometimes it was "fixed air" and sometimes the very different "dephlogisticated air"; but they did not know that the former—carbon dioxide—was a compound, or that the latter—oxygen—was an element. Priestley long thought that "fixed air" was elementary and existed in both common air and in his oxygen—his "dephlogisticated air." Many acids were known, but their components were not recognised and they were often regarded as modifications of one fundamental acid. For the chemist of this time there were all these counters, capable of being shifted and shuffled together, and nobody knew how to play with them. So many confusions existed that chemistry was building up strange mythical constitutions for its various substances. It is possible that so long as this anarchy existed any purely doctrinal statement of what a chemical element ought to be (such as that put forward by Boyle) was bound to be ineffective and beside the point.

At this moment there emerged one of those men who can stand above the whole scene, look at the confused pieces of the jig-saw puzzle and see a way of turning them into a pattern. He was Lavoisier, and it is difficult not to believe that he towers above all the rest and belongs to the small group of giants who have the highest place in the story of the scientific revolution. In 1772, when he was twenty-eight, he surveyed the whole history of the modern study of gases and said that what had hitherto been done was like the separate pieces of a great chain which required a monumental body of directed experiments to bring them into unity. He set out to make a complete study of the air that is liberated from substances and that combines with them; and he declared in advance that this work seemed to him to be "destined to bring about a revolution in physics and in chemistry." Two years later he made a more detailed historical survey of what had been done, and added experiments and arguments of his own to show that when metals are calcined they take an "elastic fluid" out of the air, though he was still confused concerning the question whether the gas which was produced on any given occasion was "fixed air" (carbon dioxide) or oxygen. He came to feel that it was not the whole of the air, but a particular gas in the air, which entered into the processes of combustion and calcination; and that what was called "fixed air" had a complicated origin—when you heated red lead and charcoal together, he said, the gas did not arise from either of the substances alone, but took something from both, and therefore had the character of a chemical compound. On the other hand, he soon came to the conclusion that the red lead when heated in isolation produced a gas which was closely connected with common air.

When he heard that Priestley had isolated a gas in which a candle would burn better than in common air, his mind quickly jumped to the possibility of a grand synthesis. Quite unjustly, he tried to steal the credit for the discovery for himself, but it is true that he was the person who recognised the significance of the achievement and brought out its astonishing implications. In April 1775 he produced a famous paper *On the Nature of the Principle that combines with Metals in Calcination and that increases their Weight,* in which he threw overboard his earlier view that the principle might be "fixed air"—carbon dioxide—and came to the conclusion that it was the purest part of the air we breathe. The idea now came to him that "fixed air" was a compound—a combination of common air with charcoal—and he soon arrived at the thesis that it was charcoal plus the "eminently respirable part of the air." Next, he decided that common air consisted of two "elastic fluids," one of which was this eminently respirable part. Further than this, he decided that all acids were formed by the combination of non-metallic substances with "eminently respirable air," so he described this latter as the acidifying principle, or the *principe oxygine.* As a result of this theory oxygen acquired the name which it now possesses, and in the mind of Lavoisier it ranked as an irreducible element, save that it contained "caloric," which was the principle of heat.

Lavoisier was not one of those men who are ingenious in experimental devices, but he seized upon the work of his contemporaries and the hints that were scattered over a century of chemical history, and used them to some purpose. Occasionally his experimental results were not as accurate as he pretended, or he put out hunches before he had clinched the proof of them, or he relied on points that had really been established by others. If he used the word "phlogiston," he soon did his structural thinking as though no such thing existed, and he disliked the doctrine before he knew enough to overthrow it. In 1783 he came out with his formal attack on the phlogiston theory in general. When a calx was reduced with charcoal he demonstrated that the transpositions of the various ingredients could be accounted for without leaving room for any passage of phlogiston out of the charcoal into the recovered metal. The French chemist Macquer had suggested in the meantime—in 1778—that phlogiston was the pure matter of light and heat, but Lavoisier ridiculed this and showed that it had nothing save the name in common with the phlogiston theory, which had reference to a solid substance possessing weight. He demonstrated that in any case the ideas of Macquer led to inconsistencies. The quarrel over the phlogiston theory seems to have aroused, as Priestley said, more "zeal and emulation" than anything else "in all the history of philosophy." At first the physicians and mathematicians in France inclined to Lavoisier, while the chemists retained their professional prejudices, and it seems to have been very much a new generation of chemists in that country who carried the victory for the new theory. In England the resistance was stronger and Cavendish refused to surrender, though he withdrew from the controversy later; Joseph Black went over to Lavoisier very late in the day; while Priestley held out, publishing in 1800 his *Doctrine of Phlogiston Established and the Composition of Water Refuted.* Like the controversy between Newton and Descartes, the new scientific issue produced something like a national division. Priestley showed an amazing liveliness and ingenuity, possessing the kind of mind which quickly seized on the importance of "fixed air" for the commercial production of mineral waters, and of oxygen for medical purposes, but could not clear the board and redistribute all the pieces on it so as to clarify the situation. At the same time his resistance to Lavoisier seems to have compelled the latter to reconsider questions and to develop his views in a more impressive way. It is curious to note that even

Lavoisier retained a shadow of the old views of combustion. Men had long been puzzled by the departure of heat and the radiation of light, and to explain these he introduced the idea of a weightless *caloric* which was involved in the process of combustion. But this proved to be easily detachable from his system later.

In 1776 Volta was firing gases with electric sparks, and he passed the discovery on to Priestley, who came to regard even electricity as phlogiston. In 1781 Priestley was exploding hydrogen and oxygen in this way—what he called a "random experiment"—and noticed that the inside of the glass vessels "became dewy." Scientists had been so accustomed to deposits of moisture from the atmosphere or to collecting gases over water, that this kind of thing had often been observed but had passed without notice, and a friend of Priestley's called Waltire repeated the experiment, but was more interested in what was really a slip which made him think that there had been a loss of ponderable heat. Cavendish confirmed the production of dew, and showed that it was plain water, that the gases combined in certain proportions to produce nothing but water, and that no weight had been lost in the course of the proceedings. It was difficult for people to believe at this time that there could not be any transmission or diffusion of weight during such an experiment, but Cavendish denied that any such loss took place. It was still more difficult for anybody to believe that water was not an irreducible element. Cavendish came to the conclusion that hydrogen must be water deprived of its phlogiston and oxygen must be phlogisticated water. Once again Lavoisier was the first to understand the situation, after learning of Cavendish's experiment, and once again he pretended to have made the actual discovery. In November 1783 he showed that water was not, properly speaking, an element, but could be decomposed and recombined, and this gave him new weapons against the phlogiston theory. He himself might have discovered the composition of water earlier than the others, but he had been unable in these years to escape from the tyranny of a preconception of his own—the view that oxygen was the great acidifying principle—which led him to look for an acid product while burning hydrogen at a jet.

He was remarkable in other ways. Finding that organic substances gave mainly fixed air and water when they burned, and knowing that fixed air was a compound of carbon and oxygen, he decided that organic substances must be largely composed of carbon, hydrogen and oxygen, and he did much towards their analysis as far as these ingredients were concerned. Already another Frenchman, de Morveau, had been striving for a revision of chemical nomenclature, and from 1782 Lavoisier worked in co-operation with him, producing a new language of chemistry which is still the basis of the language used today. The chemical revolution which he had set out to achieve was incorporated in the new terminology, as well as in a new treatise on chemistry which he wrote; and at the same time he was able to establish at last the ideas which Boyle had foreshadowed on the subject of a chemical element. He was willing in practice to accept a substance as elementary so long as it resisted chemical analysis. Over a broad field, therefore, he made good his victory, so that he stands as the founder of the modern science.

Chapter 12

Ideas of Progress and Ideas of Evolution

It is possible that the men of the Renaissance were less capable of seeing history as the ascent of the human race, or envisaging the successive centuries as an advancing series, than even their medieval predecessors had been. The men of the Renaissance were in a peculiar situation for commenting on the course of human history—their outlook highly conditioned by the unusual platform from which they turned to take their retrospect. What they saw behind them in the far distance were the peaks of classical antiquity, representing the summit of human reason, the heights which had been reached by the Greeks and since lost, the ideal for the return of which they themselves were engaging their finest endeavours. Between classical antiquity and their own time was the darkness of that medieval age which had lost contact with the legacy of the ancient world and had come to represent in their minds only a fall into error and superstition. Even if their own situation had not been so vivid, the classical thought which had such great authority for them provided a picture of the process of things in time—a theory concerning the way in which things happened in history—far removed from anything like the modern idea of progress. When they cast their minds over the whole course of centuries they were governed by the terms of this ancient outlook which at one level represented a static view of the course of things in general, and at another level (and as it regarded the internal processes within particular states or civilisations) involved a theory of decadence—the whole combining to produce in one sense change, and in another sense changelessness, under a system that might be described as cyclic.

This antique-modern view which became current at the Renaissance found explicit statement, in one of its extreme forms, in the writings of Machiavelli. Human beings, on this view, are acting throughout the centuries on the unchanging stage of the earth—the whole of nature providing a permanent scene upon which the human drama is superimposed. The human beings themselves are always alike, always made out of the same lump of dough; or, rather, we might say, they are varying mixtures of the constant ingredients of passion, affection and desire. The texture of historical narrative, therefore, would be fundamentally the same whatever the period under consideration, and to any person taking a bird's-eye view, the total appearance of the world would be very similar in all ages. One city or state might be found to be flourishing in one century and different ones at other periods, but the world in general would present the same picture—indeed, Machiavelli, for his part, explicitly tells us that he inclines to the view that the total amount of virtue in the world is always the same. At one time this virtue might be heaped together in the Roman Empire. At another time it might be scattered more thinly over the whole surface of the globe. In a fundamental sense, however, the world was forever the same.

Within any city or state or civilisation, on the other hand, the natural operation of time was to produce internal corruption; the ordinary expected routine thing was a process of decadence. This could even be observed in a parallel manner in the physical world, where bodies tended to decompose and the finest fabrics in nature would suffer putrefaction. In fact, the current science chimed in with the current view of nature, for in both these realms it was held that compound bodies had a natural tendency to disintegrate. This did not mean that all history was a long, unbroken process of decline, however; it merely meant that a rise was an extraordinary thing, somewhat against nature, and even stability for any great period—even a long resistance to the process of corruption—was a considerable feat. It was plain to everybody, and it was easily comprehensible, that if a people made a superlative endeavour, or if they were bountifully assisted by fortune —if, for example, they were endowed with a leader of special genius—they might be brought to the top of the world by a wonderfully rapid process. Only, when fortune ceased to be so extravagantly kind, or the genius died, or the unusual effort and straining were relaxed—in other words, when life returned to its hum-drum level—the ordinary tendencies of nature would begin their operation again and the normal process of decline would set in once more. Of course, if they had been pressed, many of the people who held these views concerning the general course of things would have admitted something like progress in the early stages of human history—since the days before the discovery of fire, for example. But it does not seem that their broader views concerning the succession of ages were governed by facts like these.

On this view of the universe, time and the course of history were not considered to be actually generative of anything. On this view also, one had no conception of a world opening out to ever grander things, to an expanding future—there was not even an idea of a civilisation that was supposed to develop indefinitely. Men assumed rather the existence of a closed culture, assumed that there were limits to human achievement, the horizon reaching only to the design of recapturing the wisdom of antiquity, as though one could do no more than hope to be as wise as the Greeks or as politic as the Romans. On the same view the notion of something like a "Renaissance" was a comprehensible thing, associated in a way with ideas arising out of the fable of the phœnix; and some signs of such a notion are visible in the later middle ages, when the humanist movement was associated for a time with the dream of rescuing the papacy from Avignon and the empire from Germany, the wheel coming into full cycle as men looked forward to the renewed supremacy of Rome.

The reassertion of these ancient ideas on the subject of the historical process helps to explain why at the Renaissance it was almost less possible to believe in what we call progress than it had been in the middle ages. If anything it was more easy to believe in something of this sort in the realm of spiritual matters than in any other sphere—to believe in stages of time succeeding one another in an ascending series (though possibly still by sudden jumps) and so to find meaning and purpose in the passage of time itself. The transition from the Old Testament to the New, and the notion of a Kingdom of the Father, succeeded by a Kingdom of the Son, with a Kingdom of the Spirit to follow, were examples of this. It has been suggested that the modern idea of progress owes something to the fact that Christianity had provided a meaning for history and a grand purpose to which the whole of creation moved. In other words, the idea of progress represented the secularisation of an attitude, initially religious, which looked to a fine fulfilment in some future, far-off event, and saw history, therefore, as definitely leading to something.

Most of the basic ideas in the Renaissance view of history are still clearly present in the controversies of the latter part of the seventeenth century; but the famous quarrel between the Ancients and the Moderns—the controversy in the course of which a more modern view of progress was hammered out—is already visible at the time of the Renaissance. In this earlier period, however, there was a sense in which everything was on the side of the Ancients—and possibly rightly so, since antiquity had still many things to teach to western Europe—so that the real issue was the question of the degree to which the discipleship ought to be carried. Machiavelli said that the Romans ought to be imitated in every detail, and he was reproached for disparaging gunpowder because the Romans had not used it. Guicciardini, however, insisted that a more fluid and elastic policy of imitation was necessary, because conditions had changed. He thought that, when taking one's bearings for the development of a military science, one ought to have all the resources of modern invention before one's mind. We have already seen that one school of Renaissance students went on taking the medical teaching of the ancient world as it had been transmitted by the Arabs; another school would

Renaissances

antiquity vs. modern

be content with nothing short of the ancient Greek text itself, and a purer knowledge of antiquity.

Even in the sixteenth century, however, we meet more than once a fact which has particular significance in connection with that development in ideas which we are considering. Some people realise that the mariner's compass, the printing of books and the use of artillery represent achievements as momentous as anything produced in the world of antiquity. Some people even bring out this argument without any hint of the influence which the Far East may have had on these developments—they simply use these things as demonstrations of the prowess of the West, the accomplishments of the Moderns. The new worlds opened up by geographical discovery and the multitude of published books are calculated to become a heavy counterweight to the much-vaunted superiority of the Ancients.

It could not be very long before it was realised that certain forms of scientific knowledge gained something by the very lapse of time, whether by the accumulation of data—the sheer increasing aggregation of observed facts—or by the continual revision of the results and the improvements in actual method. This had been particularly noticeable hitherto in the case of astronomy. Before the end of the sixteenth century Giordano Bruno was pointing out that even in the ancient world Ptolemy had built on the observations of his predecessors; these in turn had had the advantage of starting from the achievements of others earlier still; while Copernicus, collecting all that his predecessors had done, was in a better position than any of them to know the situation of things in the sky. In fact, said Bruno, it is we who are the Ancients, and who enjoy the benefits of the accumulated experience of mankind; and the age of classical Greece belongs rather to the childhood of the world. The argument that we are the more ancient appears on a number of occasions in the seventeenth century; but the comparison of the whole of history with the life of a man was capable of being used to the opposite purpose, and it is possible that Francis Bacon gave a wrong impression when he showed that the Moderns were really the seniors; for it became necessary before the end of the seventeenth century to deal with an extremer version of the doctrine of decadence—the view that this was the old age of the world and that nature was unable to put out the same powers as before—a notion which involved, not a static view of the total process of things in time, but a feeling that nature herself was suffering from a long process of exhaustion. It was Fontenelle who set out to answer this argument at the end of the century, and replied that if the human race was to be likened to a man, it was to a man who acquired experience without ever growing old. Though on its mechanical side the scientific revolution glorified Archimedes, while the revived forms of atomic philosophy called attention to another aspect of the thought of the ancient world, the conscious demand for a new science and a new outlook, the discrediting of Aristotle and the insistence by Descartes on the importance of unloading the mind of all its traditions—all those things dealt a great blow at the authority of antiquity.

It was the glamour of Versailles and the literary glory of Louis XIV's reign, however, which led to the new and more fundamental form of controversy between the Ancients and the Moderns in the latter part of the seventeenth century—an important stage in the development of the idea of progress. One aspect of the self-glorification in which the age indulged was the spread of the idea that the glories of ancient Greece had been revived by the literary giants of *le grand siècle*. The controversy was precipitated by Charles Perrault, who in 1687 published a rhymed work on *The Age of Louis the Great* and between 1688 and 1697 produced his *Parallel between the Ancients and the Moderns*. But he had been preceded earlier in the century by another writer, Desmarets, who had carried on the literary controversy against the Ancients, and had claimed, for example, that Christian subjects provided better themes for the poet than ancient mythology—a thesis which he had illustrated in epics of his own, though it was better illustrated by Milton in England. He had likened antiquity to spring, and modern times to the mature old age and, as it were, the autumn of the world; the faults of earlier centuries were now corrected, he said, and those who come latest must excel in happiness and knowledge. It is interesting to note that this man, Desmarets, had definitely bequeathed his mantle and entrusted the continuation of the controversy to Charles Perrault, and it was Perrault who, in fact, created a much greater sensation and was responsible for the crucial controversy. He claimed that Plato had been tedious on occasion, and, like others, he was prepared to maintain that even Homer had nodded at times. In his view the age of Louis XIV had excelled the literary prowess of the ancient world; for, just as the Ancients only knew the seven planets and the most remarkable stars, while we had discovered the satellites of the planets and numberless tiny stars, so the Ancients had known the passions of the soul only *en gros,* while we knew an infinity of subtle distinctions and accompanying circumstances.

It is interesting to note that, though the controversy which ensued was so essentially a literary one, the decisive fact that emerged—and the argument which proved effective on the side of the Moderns—was connected with the achievements in natural science, and in the related aspects of life and society. And by this time it is clear that what we should call a more historical attitude had made itself apparent in the discussion of the position of science in the panorama of the centuries. Previously, as we have seen, there had been an idea that a scientific revolution was necessary, but it had been thought that it would occur and complete itself as a great historical episode, putting a new view of the universe in place of the Aristotelian one; Bacon had imagined that the work of experiment and discovery could be achieved in a limited period, while Descartes had thought it important that the revolution should be carried out by a single mind. A cataclysmic view of such forms of achievement was still prevalent and it was consonant with this whole outlook that the age believed in the formation of states by a social contract, rather than as a growth taking place, so to speak, in nature. In the latter part of the seventeenth century, however, it becomes clear that men have a vision of science as a young affair with all its future before it—an ever-expanding future—and Fontenelle points out that the sciences are still in their cradle. In the new stage of the controversy between the Ancients and the Moderns it is difficult for the former to deny the progress that has taken place on this side, and though there is a tendency to make discriminations and to say that the art and literature of the ancient Greeks are still unexcelled, the modern world may come so near, even in the field of poetry, that it seems to have the advantage on the whole count. In any case, the sum of science, industry, improvements in society and the development of communications form an argument in favour of the Moderns, and a popular debating-point was used to give the Moderns the palm on the ground of the general opulence which had come to prevail. The general impression of abundance, the feeling of comparative security—of insurance against mischance or disease—the progress of luxury and the wonderful machines were described in a manner which reminds us of Macaulay in the nineteenth century; and it was noted that the citizen of Paris walked the streets in greater splendour than used to belong to a Roman triumphal march. There began to be even a certain intolerance of the barbarity of the preceding centuries—we gain an impression of modernism when we see men indignant that the streets of Paris had had to wait so long before being paved. People dreamed of the time when what they called "this mechanic civilisation" would be transported to countries hitherto uncivilised. The whole tendency of the new philosophies was to shelve the idea of Providence, which seemed a capricious interference with the laws of nature; and, indeed, the new power which was coming to be acquired over material things encouraged the idea that man could, so to speak, play Providence over himself. The new historical work—Fontenelle's study of myths, for example, and then the writings of Vico, the examination of primitive societies and the discussion of the development of human reason—encouraged the idea that men possessed natural reason, and this only required to be disengaged from traditions, institutions and mal-education. A general improvement was possible in individuals themselves, then, and was already observable at that very time. The way was open for the doctrine of the perfectibility of man, a perfectibility that was to be achieved by remedying institutions.

The transition to the idea of progress was not one that could be completed in a single simple stage, however, and at the close of the seventeenth century we can neither say that the idea had been fully developed nor feel that its implications had become generalised. Even the advocates of the Moderns against the Ancients could hardly be described as the apostles of what we mean by progress. Even Perrault, though he thought that civilisation had come to a new peak in the France of Louis XIV, did not consider that the ascent would be prolonged indefinitely, but held that when the present epoch had had its run the world would return to normal, so that the process of decline would soon start over again. Perrault, in fact, was of the opinion that there would not be many things for which the France of Louis XIV would need to envy posterity. And Fontenelle, though he was conscious of the widening vistas which the future promised to the natural sciences, was too well aware of the limitations of human nature to share the illusions of many of the *philosophes* concerning the general improvement of the world. What is asserted in the controversies at the close of the seventeenth century is the fact that nature is the same in all ages—she can still put forth men of genius who are capable of holding their own with the giants of ancient times. Fontenelle sets out to show that nature in the seventeenth century has lost none of her prolific power—the modern oak-trees are as big as those of ancient Greece. At the same time—perhaps almost incidentally—the idea is asserting itself that a general improvement is taking place in conditions, and particularly in the things that concern the welfare of ordinary human beings. We can even discern that these are the things which are coming to bulk most largely in the public mind, the sheer weight of them serving to turn the scale in favour of the Moderns. Writers were able to address themselves to this

idea of a general progress in human conditions as though it were a matter generally understood.

Even in the eighteenth century certain of the prevailing ideas or prejudices are awkward to reconcile with any scheme of history on the basis of progress. The regard for native reason, and the view that this was liable to be perverted by institutions, led to a certain amount of day-dreaming about the "noble savage" and the evils of civilisation itself, as is illustrated in the writings of Rousseau. John Wesley, when he went to America in his youth, not only had the idea of carrying out a missionary task, but thought that light would be thrown on the Bible itself if native minds, uncorrupted by centuries of commentating, could suddenly be confronted with the scriptural revelation. The political ideas of eighteenth-century England are formulated with reference to a scheme of history which still assumed the golden age of the constitution to have existed somewhere in the distant past. The literature of parliamentary reform in the 1770s and of the Yorkshire Association of 1780 stresses the fact that annual parliaments and universal manhood suffrage prevailed in Anglo-Saxon England, though subsequent tyrants had tried to destroy the evidence for the existence of our ancient liberties. Embedded in the whiggism of the eighteenth century are maxims and theses taken straight from Machiavelli concerning the tendency of liberty to decline if men do not strain their ingenuity in every generation to preserve it. At the most primitive stage in our history we had had the best of constitutions; but in the reign of George III the corruption of this was regarded by some men as having become almost complete.

In fact, the attempt to embrace the whole course of things in time and to relate the successive epochs to one another—the transition to the view that time is actually aiming at something, that temporal succession has meaning and that the passage of ages is generative—was greatly influenced by the fact that the survey became wider than that of human history, and the mind gradually came to see geology, pre-history and history in due succession to one another. The new science and the new history joined hands and each acquired a new power as a result of their mutual reinforcement. The idea of progress itself gained additional implications when there gradually emerged a wider idea of evolution. It may be useful to compile—though necessarily at second-hand, and though the time has perhaps not yet come for any profound understanding of the matter—a sketch of the developments which took place in this field in the course of our period.

STOP

The history of the idea of evolution is connected with the development of systems of classification in the realm of plants and animals. A rapid survey of this story might start with the work of John Ray in the closing years of the seventeenth century; for Ray seems to have discussed the notion of "species" more than any of his predecessors, and some of his remarks support the traditional view which regarded the various species as having been absolutely fixed since the day when God rested from the work of Creation. This latter view was adopted by the Swede, Linnæus, whose work as a classifier in the 1730s gave him a remarkable reputation that lasted down to the time of Darwin. He assumed that all individuals in a given species could be traced back to an original pair produced at the Creation, and it was to be significant that he gave the weight of his great authority to the idea of the immutability of species. It happens at times that it is not the best of a man's thoughts which are remembered and which gain currency by association with his name. Linnæus, towards the end of his life, became more careful in his discussion of the borderlines between the species—partly because he had discovered much overlapping, and partly because he had done much in the way of hybridising in his own garden. It does not appear to have been this side of his work which influenced the world, however. It transpires, in fact, that his successors were more rigid in their ideas on this subject than he himself had been in his better moments.

Before his time the famous German philosopher, Leibnitz, had adopted a more flexible attitude. At the beginning of the eighteenth century he had emphasised the continuity of creation and the unbroken gradation of organisms in nature. He had called attention to fishes which had wings and could live out of water, as well as to birds which inhabited the water and had cold blood like fish; and, similarly, he had noted the case of animals that came near to being birds. His view of nature appears to have been influenced by the achievements effected with the microscope in the latter part of the seventeenth century, when even a drop of water had been shown to be teeming with life. Instead of reducing the universe to atoms which were rigid and lifeless, he regarded it as composed of small particles of matter which were living monads or life-principles, and which served as the basis of living organisms—a view which influenced a number of biologists, especially as it made it more easy to explain the variety of combinations that existed in nature and to account for the origins of various forms of life without a special act of creation. The assumption that these minute living particles ultimately accounted for the various forms of life that existed in the world helped to prolong the idea of spontaneous generation into a period which would otherwise have hardly found it plausible. John Locke had pointed out how vague were the boundaries between the species, all descending "by easy steps and a continued series of things that in each remove differ very little from one another down to the lowest and most unorganical parts of matter." He held that the various species had no separate existence in reality—they were created, in fact, by the mind of man, reducing nature to order, and were not the work of nature herself. Altogether, in the days of Locke, it seems to have been rather more easy to question the immutability of species than during most of the eighteenth century. The influence of Linnæus would seem to have been partly responsible for a certain hardening of ideas that took place in the latter period.

The eighteenth century indeed brings to a climax in scientific thought, in philosophy and in literature, the notion of "the Great Chain of Being"—the idea of an infinitely graded series of creatures stretching down to inanimate nature and up to God Himself, with man standing somewhere near the middle, if as high as that. The entire range of the series was not regarded as necessarily existing on our particular planet; but as many varieties, as many individuals, were presumed to exist as could consistently co-exist, so that all the potentialities of Being should be actualised and—even if at the cost of accompanying evil—all the possibilities of good should be brought to realisation. Each individual in the series existed for its own sake, and not merely as a link in the chain—certainly not for the mere purpose of serving man as the chief end of creation. But any known gradation might be imagined as capable of further sub-division, and interest particularly concentrated itself on the missing links, especially at the transitions between plants and animals and between animals and man. Hence the excitement when in 1739 Trembley rediscovered the fresh-water polyp *Hydra,* which seemed to provide the missing link between vegetable and animal life; also the interest in the Hottentots who from the late seventeenth century seemed almost to represent an intermediate stage between the ape and *homo sapiens.* The whole notion of this "chain of being" could be combined, however, with the idea of the immutability of nature; for all the potentialities of being might be regarded as always existing at the same time in order that the fullness of the universe should be complete. It was even inconvenient to learn that from fossils one might infer the existence in the distant past of species now extinct. And though early in the eighteenth century one meets the suggestion that the earliest animal forms were those that lived in the sea, one finds also the theory that all future generations existed inside the first—all future generations existed, pre-formed, in the first individual—200,000 million miniature men in the original Eve—so that time had nothing to do with the shaping of them. The remarkable development in the course of the eighteenth century was the transformation of the "Chain of Being" into historical terms, into a ladder by means of which the living world had come into its present state. Even in a wider sense than this, the universe was coming to be seen not merely as existing spatially but also as possessing a history.

Although fossils had long been associated with the sea and the Flood, more fantastic conjectures were common, and there was a view that they had grown from seeds carried by underground passages to the tops of mountains and that these had been fertilised by the snow. Alternatively it was suggested that before earth and water had been separated at the Creation, they had existed in the primeval mud, and when this had been pressed like a sponge the birds and fishes had run out with the water, while plants and animals had been drawn on to the land, some creatures, however, missing their emancipation owing to accident. There was even an idea that the fossils on Mont Cenis had been dropped from the food of pilgrims passing that way—fish, for example, that had become petrified in the course of time. As the eighteenth century proceeded, however, it came to be accepted that fossils were formed by deposits which had been made by the ocean in prehistoric times. In the days before modern specialisation, the collectors of fossils seem to have often been men who had also been caught up by the antiquarian movement. The scientific revolution combined with the parallel development of history, and one now tended to envisage the world as existing and developing through the succession of ages.

Already the minds of certain people had ranged more widely over the spectacle of the earth in time, and behind the historical realm, behind the story of the animal and vegetable kingdoms, had gained an impression of a drama more truly primeval, realising that there was another tale to be told in terms of geological epochs. Interest in all kinds of speculation concerning the history of fossils and rocks was becoming particularly lively again in the latter part of the seventeenth century. Leibnitz had the idea that this globe had once been a sun, and gives a picture of the mother earth solidifying so that its encrusted surface crinkled into mountain-ranges. The materials were being assembled for a larger apprehension of the whole process of things in time—transposing on to the canvas of all the ages that width of survey, that comprehensive vision, which had already done so much to reveal the earth's position in space. In the most ancient days there had existed vague ideas concerning the evolution of all the things in the world from a sort of primeval slime, or the emergence of gods from the hatching of primordial eggs. The introduction in modern times of a view which envisaged the whole universe in terms of historical process was a new thing, however, and represents an important stage in the development of the modern mind. The transition which took place in the eighteenth century would have implied a radical change in the human outlook even if it had not brought the scientists into conflict with the Biblical story of creation. And though we must regard the idea of evolution—like the idea of progress—as only imperfectly achieved at the end of that century, it would seem to be true that almost all the essential ingredients of Charles Darwin's system had made their appearance by that date. Science and history had come together to present the idea of the whole of nature advancing slowly but relentlessly to some high goal.

Georges Louis Leclerc de Buffon produced in the fifty years from 1749 an *Histoire Naturelle* which was to stand as one of the signal products of eighteenth-century science. His industry was great, but his researches were not original or profound, and he was over-hasty in generalisation, fulfilling in part the rôle of a populariser, and labouring greatly over his style (though the style was too pompous and turgid) while apparently desirous that his contribution to science should serve also the sentimental education of men. He attempted to see nature as a whole, produced a vast synthesis and sought to give a picture of the history of the earth, regarded as a habitation for living creatures. Indeed, if Newton had appeared to reduce the inanimate world to a system of law, Buffon would seem to have set his mind on a parallel achievement, and even a wider one—comprising in his synthesis biological phenomena and expanding into the realm of history. His views were not always the same and he has been accused of hovering between Biblical Creation and the ideas of Evolution, or of an unwillingness to challenge too seriously the teaching of the Church.

Along with Leibnitz he believed that the earth had once been in an incandescent state; and it was his view that, like the other planets, it had actually been part of the sun, but had broken away after a collision with a comet. He rejected the tradition that this globe was only six thousand years old and made an attempt to set out the periods or stages of its history; a time when the mountain-ranges were formed; another time when the waters entirely covered the face of the globe; a period when the volcanoes began to be active; a time when tropical animals inhabited the northern hemisphere; and a time when the continents came to be separated from one another. He held that changes took place in the vegetable and animal kingdoms as the earth passed from one of these epochs to another. In connection with this problem he made it part of his task to study the fossils which were abundant in the stone that was then being used for building in the city of Paris.

We find him making surmises concerning the origins of the various forms of life and the place of their appearance, as well as the subsequent influence of physical conditions in general upon them; though here, as elsewhere, he is not always consistent. He was prepared to think that a special act of creation was not necessary to account for the emergence of living creatures on this planet, regarding life as, so to speak, a quality or potentiality of matter itself. He held something like Leibnitz's idea that every plant and animal was composed of a mass of minute particles, each of which was a pattern of the whole individual; and this enabled him to explain the origin of living creatures without reference to an act of creation. He tried to show that no absolutely definite boundary existed between the animal and the vegetable kingdoms. Nature always proceeds by *nuances,* he said. "It is possible to descend by almost insensible degrees from the most perfect creature to the most formless matter." There are many "intermediate species," many things that are "half in one class and half in another." For these reasons he represented a reaction against the orthodoxy of the Linnæan tradition, and against the rigidity of those writers who tended to regard mere classification as an end in itself. He could deny the view that the species were immutable —eternally separate from one another. He had misgivings when he considered the infertility of hybrids, however, for this suggested that the species were real entities, each insulated against the rest.

In some respects his ideas would be strange to modern believers in evolution; for he seems to have believed that various living species were degenerate forms of types which had once been perfect. He put forward the view, however, that the environment directly modified the structure of animals and plants, and he implied that the acquired characteristics were transmissible. He declared that many of the species "being perfected or degenerated by great changes in land and sea, by the favours or disfavours of Nature, by food, by the prolonged influences of climate, contrary or favourable, are no longer what they were." He hinted at the possibility of a common ancestor of the horse and the ass, and even said that he would have extended this supposition to the case of man himself if the Bible had not taught the contrary. Of the orang-outang he declared: "If we do not take the soul into account he lacks nothing that we possess." He wrote: "The pig does not appear to have been formed upon an original special and perfect plan, since it is a compound of other animals: it has evidently useless parts, or rather parts of which it cannot make any use, toes all the bones of which are perfectly formed and which nevertheless are of no service to it." He argued that since some organs in living creatures no longer seemed to have any utility, times must have very radically changed. He had some conception of a struggle for existence which eliminated the unfit and preserved the balance of nature.

It is clear that we are already in a transformed intellectual world. Buffon's work envisages the universe throughout the ages and shows a remarkable sense for the continuous flow of things in time. It involves a new conception of the relations between man and nature, revealing a willingness to study man in his aspect as part of nature. Henceforward the idea was spread that many features in the anatomy of man pointed to his having descended from quadrupeds, and, indeed, to his not being fully adapted to an erect position even now. In the same period La Mettrie, in a series of bold speculations, based on the idea of living particles which were supposed to come together to form live individuals, had discussed the emergence of living creatures on the surface of the earth, and had found a natural explanation for the origin of man.

It would seem, however, that apart from great syntheses and famous names—apart from men like Buffon—the accumulated work of a multitude of famous enquirers contributed to the developments which were taking place. In the latter half of the eighteenth century researches in varied fields were preparing the way for a more solid advance towards the modern idea of evolution. Caspar Friedrich Wolff in Germany would seem to have played his part when he made a comparative study of plant and animal development and pointed out the cell-tissue which was common to both. In 1759 and 1768 he attacked the popular pre-formation theory which regarded the female as containing the germs of all the generations to follow, one incapsulated within another, and each comprising the pre-formed individual, already supposed to exist in miniature. He showed that limbs and organs in the embryo passed through various successive shapes, and this process seemed to him to exhibit the operation of a kind of life-force which worked upon simple homogeneous organic matter, building it up into structures. Koelreuter studied pollen and pointed out the importance, on the one hand of insects, on the other hand of the wind, in the fertilisation of flowers. Conducting experiments in hybrid formations, he showed, for example, that when hybrids were mated with their parent-species a reversion would take place—the characteristics of the parent would return. Christian Conrad Sprengel showed how certain flowers required certain forms of insect for their fertilisation, while others might be fertilised by insects of varying sorts; and the position of the nectaries in each flower would be adapted to the insects which were accustomed to servicing that flower. Petrus Camper in Holland studied faces and stressed the differences between those of human beings and those of apes which might be regarded as most resembling man himself—a topic which provoked a good deal of controversy in the latter half of the eighteenth century. It would appear that the old views concerning the immutability of species were liable to be modified by men who had interested themselves in hybridisation, or, like Erasmus Darwin, in the breeding of horses, dogs and sheep.

The transition to evolutionary thinking was assisted not only by the spread of the historical outlook and the idea of progress, but also by philosophical tendencies, such as the inclination to see the whole world as a living thing, to believe in an *élan vital* and to postulate some spiritual formative principle, operating throughout nature and gradually realising itself. Jean Baptiste Robinet (1735–1820) shows the effect of these ideas in his treatise on Nature which appeared between 1761 and 1768. He sees all organic creatures in a linear scale, but to him all the lower forms of life are an adumbration of the future figure of man, and even in the early stages of the world's development he looks for the suggestion of the human form. In his view, the lower creations were a necessary intermediate stage before man himself could be produced as the crown of creation. All the parts of the human form had to be tried out in every conceivable combination before the shape of a man could really be discovered. The history of the earth itself was "the apprenticeship of Nature in learning to make a man." Robinet illustrates another view which was current in this period, and which assisted evolutionary speculation—the idea that the atoms out of which everything was composed were not merely dead matter but were individually possessed of both life and soul. Inorganic matter could build itself up into combinations that formed living creatures—there was no real gulf between the animate and inanimate. There were infinitely small gradations between all the things in nature, and the chain of being was a continuous one—a single pattern at the basis of everything, a single prototype, was discoverable behind all the variations and provided the continuity. And this idea, which is found in Robinet, was to receive further development in the work of Herder and Goethe in Germany. Furthermore, an age that had long been familiar with discussion about the influence of climate and environment on the various sections of the human race, was ready to reflect—as Robinet did—upon the way in which the external world might condition the development of plants and animals.

Also geological ideas and hypotheses, which had been appearing for a long time, were beginning to shape themselves into a science, and from about the year 1775 acquired more importance than ever before. By the end of the eighteenth century a handful of sciences on which geology itself depends had at last been brought to a state of ripeness. The view that all rocks had been precipitated in a primordial ocean or from the fluid that formed the original chaos is one that goes back to antiquity. It was challenged in the 1740s by a theory which postulated the volcanic origin of all rocks; and from the 1760s the Neptunists were in conflict with what were called the Vulcanists or Plutonists. In 1775 Werner in Germany discussed the earth's surface in a more systematic manner than his predecessors, and he maintained the aqueous origin of rocks—the view that was then the more popular. James Hutton, who wrote in 1788 and 1795, asserted the igneous origin of rocks and rejected the idea that the earth had acquired its formation through a series of great catastrophes. He preferred to interpret the past in the light of the known present and sought to account for the earth's present condition by reference to processes still observable, forces still in operation and principles already familiar. The Catastrophic theory was met by a doctrine of Uniformitarianism, and Hutton, though he had little influence in his own day, marked out the path for the future development of geology.

Charles Bonnet (1720–93) brought his religious belief to support the prophetic faith that he had in the progress of the world and the advancement of nature. He too held the view that the final units out of which everything in the world is built are living, indestructible things, as old as the universe, and each of them primarily a "soul." He saw the whole of nature in a linear series, running from the simple to the complex, but each member differing only infinitesimally from its neighbours, so that continuity was unbroken from the mineral to the vegetable world, and then till the animal kingdom was reached, and finally until the emergence of man. Bonnet was particularly interested in the transitional forms such as the flying fish, the bat, the polypus and the sensitive plant—particularly interested in the orang-outang, which he said was susceptible of being educated into a polite and creditable *valet de chambre.* He was prepared to believe that evolution had been carried further on some plants than on the earth—the stones having sensations, the dogs being capable of intellectual exchange, and men achieving the virtue of angels. But he held the pre-formation theory—held that the first female holds inside her the germ of all succeeding generations—and not merely the germ but the miniature form of the adult individuals. According to Bonnet the world was periodically engulfed in great catastrophes of which the last was the great Flood that we associate with Noah. In such catastrophes the bodies of all living creatures were destroyed but the germs of the future generations lived on, and after the catastrophe they led to creatures that were higher in the scale of being.

The eighteenth-century part of the story really came to its climax in two men whose important writings appear at the opening of the nineteenth century, at a time when Paris had become the chief centre of biological study—Jean Baptiste de Monet (who is known as Lamarck) and Georges Léopold Cuvier. It was fortunate that they lived "in the Paris basin, a vast cemetery of corals, shells and mammals; and not far from extensive deposits of cretaceous rocks packed with fossil invertebrates." They became respectively important as the virtual founders of invertebrate and vertebrate palæontology. Lamarck was a man of brilliant intuitions, but his speculations sometimes ran too far ahead of his scientific achievements, and perhaps it was partly for this reason that he secured hardly any following in his own day. He began by thinking that the species were fixed, but he came over to the view that "in reality only individuals exist in nature." He began by arranging the zoological groups in a vertical scale, but as time went on he allowed the rungs of the ladder to spread wider horizontally, till the system became more like a genealogical tree. He did not believe in the continuous Chain of Being, but thought that there were leaps in nature—a gap for example between the mineral and the vegetable realms. He held, however, that life was generated spontaneously from gelatinous or mucilaginous matter, the process being assisted by heat and electricity. He rejected the view that the history of the animal world could be explained only by a series of colossal universal catastrophes which changed the whole distribution of land and sea. He believed that the earth had had a slower and more continuous history and that only in a much more gradual way had the extinct species revealed in fossils been turned into the ones which inhabit the world today. He had an impressive sense of the the colossal length of geological time, and saw animal life as continuous—without total extinctions and total renewals in the times of cataclysm—though more gradual changes in the world altered the environment of living creatures. Changes in environment did not operate directly to alter the species in the course of time; they acted through the nervous system on the whole structure of creatures, and an urge inside the creature itself had its part to play in the evolutionary process. Altered wants led to altered habits; and the organs of animals were enlarged or diminished according to the degree to which they were put to use. Lamarck held that it was not the form or character of the living body itself which decided the habits of the creature in question. It was rather the habits and manner of life which decided the fashioning of the organs—moles and blind mice having lost their sight because they had lived underground for many generations, while swimming birds had acquired webbed feet because they stretched out their toes in the water. If new characteristics were acquired by both sexes Lamarck assumed that they were transmissible; and he thought that if a number of children were deprived of their left eye at birth, only a few generations would be necessary to produce a race of one-eyed human beings. He was unfairly charged with the theory that animals could create new organs for themselves by merely wanting them. Although he spoke as though everything were the product of blind mechanical forces, and believed that even the soul was only the product of such forces, his theory ascribed a rôle to a certain urge within the individual creature, an urge which converted itself into an active fluid, flowing into the channels required, as in the case of the giraffe which strained its neck to reach the high branches of trees. Because he believed in this urge, and in a kind of aspiration of living things to attain greater complexity, as well as thinking that, up to a certain limit, life itself tends to increase the volume of any body (or part of a body) possessing it, some people have been able to argue that he shared some of the "vitalistic" notions of his time.

His contemporary, Cuvier, made a greater impression upon the world at the time, and has been described as "the first man to enjoy the full luxury of a bird's-eye view of the whole of life spread out backwards in time as well as around in space." He held that great catastrophes which had affected the nature of the earth's surface had altered the character of the animal world at different periods. The view seemed to him to explain those cases where stratifications now discovered at great heights and in an inverted position were shown by fossils to have taken place in the sea. It has been suggested that he was over-influenced by the fact that French geological research had been so largely concerned with the Alps, where subverted formations of this kind had aroused considerable interest. Cuvier's evolutionary theory did not require the long periods of time which Lamarck had had to postulate for the maintenance of continuity in a protracted course of gradual development. Also, he did not agree that the species underwent changes as a normal effect of habitat and environment—to him the changes which occurred in the animal kingdom had a catastrophic character. At any particular date the existing species were immutable in his opinion, and change was brought about when some upheaval resulted in their destruction. At each catastrophe, however, some isolated region would be spared, so making it possible that the human race itself, for example, should maintain its continuity.

He made a more methodical study than Buffon of the fossils which were so plentiful in the region of Paris. He was not content to study the separate parts of the body in isolation, but examined the way in which they would be adjusted to one another, showing how the carnivorous animal would have the right kind of teeth, jaws, claws, digestive canal, good visual organs and the power of rapid motion. This attention to the correlating of the parts enabled him to make more successful reconstructions from fossils and fragments; and he was able to show, for example, how an extinct mammoth was more closely related to the Indian elephant than the Indian and African elephants are to one another.

He went further than Lamarck in his refusal to arrange all living creatures in a single ascending or descending series, and in his insistence that the animal world should be divided into separate groups, each with a ground-plan of its own. This important idea, of which there had been a hint in the work of Charles Bonnet, implied that there was not a single evolutionary progression, but parallel lines of development in the various groups. It stood as a warning against any attempt to draw straight lines of direct comparison between highly developed and highly specialised creatures which might have run far away from one another as a result of separate developments within separate groups. The new system also enabled Cuvier to make great improvements in classification. It was a necessary step towards the establishment of a practicable theory of descent.

It has been noted that by this date all the ingredients

Suggestions for Further Reading

Aiton, E. J.—"Galileo's Theory of the Tides," *Annals of Science*, Vol. X (1954).

Andrade, E. N. de C.—*Sir Isaac Newton* (London, 1954).

Armitage, A.—*Copernicus the founder of modern astronomy* (London, 1938).

Bayon, H. P.—"William Harvey, physician and biologist" (*Annals of Science*, Vols. III and IV, London, 1938-40).

Bell, A. E.—*Christian Huygens* (London, 1947).

Bloch, Ernst—"Die antike Atomistik in der neueren Geschichte der Chemie," *Isis*, Vol. I (1913-14).

Boyer, C. B.—"Notes on Epicycles and the Ellipse from Copernicus to Lahire," *Isis*, Vol. XXXVIII (1947).

Broad, C. D.—*The philosophy of F. Bacon* (Cambridge, 1926).

Brunetière, F.—*Études critiques sur l'histoire de la littérature française*, 5me série (Paris, 1896), pp. 183-250, "La formation de l'idée de progrès au XVIIIme siècle."

Bulletin of the Polish Institute of Arts and Sciences in America, Vol. I, No. 4 (July 1943), pp. 689-763 (essays on Copernicus by A. Koyré, etc.).

Burtt, E. C.—*The metaphysical foundations of modern physical science* (London, 1925).

Bury, J. B.—*The idea of progress: an inquiry into its origin and growth* (London, 1920).

Carré, J. R.—*La philosophie de Fontenelle* (Paris, 1932).

Clark, G. N.—*Science and social welfare in the age of Newton* (Oxford, 1937).
The seventeenth century (Oxford, 1947).

Cooper, L.—*Aristotle, Galileo and the tower of Pisa* (Ithaca, New York, 1935).

Dampier, Sir William C.—*A History of Science* (3rd ed., Cambridge, 1942).

Davis, T. L.—"Boyle's conception of Elements compared with that of Lavoisier," *Isis,* Vol. XVI (1931).

Dingle, H.—"Copernicus's Work," *Polish Science and Learning* (July 1943).

Dreyer, J. L. E.—*History of the planetary systems from Thales to Kepler* (Cambridge, 1906).

Duhem, P.—*Le mouvement absolu et le mouvement relatif* (Paris, 1905).

Études sur Léonard de Vinci (Paris, 1906-13).
Le système du monde (Paris, 1954).
The Aim and Structure of Physical Theory (Eng. transl., Princeton, N. J., 1954).

Foster, Sir Michael—*Lectures on the history of physiology during the 16th, 17th and 18th centuries* (Cambridge, 1901).

Gade, J. A.—*The Life and times of Tycho Brahe* (Princeton, 1947).

Gilson, E.—*Études sur le rôle de la pensée médiévale dans la formation du système cartésienne* (Paris, 1930).

Goldbeck, E.—*Keplers Lehre von der Gravitation* (Halle a.S., 1896).

Hall, A. R.—*Ballistics in the Seventeenth Century* (Cambridge, 1952). *The Scientific Revolution, 1500-1800* (London, 1954).

Hazard, P.—*La crise de la conscience européenne* (1690-1715) (3 vols., Paris, 1935).

Jeans, Sir James—*The growth of physical science* (Cambridge, 1947).

Johnson, F. R.—*Astronomical thought in Renaissance England* (Baltimore, 1937).

"The Influence of Thomas Digges on the Progress of Modern Astronomy in Sixteenth Century England," *Osiris,* Vol. I (1936), and Larkey, Sanford V., "Thomas Digges, the Copernican System and the Idea of the Infinity of the Universe in 1576," *Huntington Library Bulletin,* No. 5 (April 1934).

"Robert Recorde's Mathematical Teaching and the Anti-Aristotelian Movement," *ibid.*, No. 7 (April 1935).

Koyré, A.—*Études galiléennes* (3 fascicules, Paris, 1939-40).

Kuhn, T. S.—"Robert Boyle and Structural Chemistry in the Seventeenth Century," *Isis,* Vol. XLIII (1952).

Lenoble, R.—*Mersenne ou la naissance du mécanisme* (Paris, 1943).

Lovejoy, A. O.—*The Great Chain of Being* (Cambridge, Mass., 1950).

Lovering, S.—*L'activité intellectuelle de l'Angleterre dans l'ancien "Mercure de France"* (Paris, 1930).

McColley, G.—"The seventeenth century Doctrine of a Plurality of Worlds," *Annals of Science,* Vol. I (1936).

McKie, D.—*Antoine Lavoisier, the father of modern chemistry* (London, 1935).

Meyerson, E.—*La déduction relativiste,* especially Ch. IV (Paris, 1925).

Identity and reality, especially Ch. III (London, 1930).

Mieli, A.—*La science arabe et son rôle dans l'évolution scientifique mondiale* (Leiden, 1939).

Nordenskiöld, N. E.—*The history of biology* (Eng. transl., London, 1929).

Ornstein [Bronfenbrenner], M.—*The rôle of scientific societies in the seventeenth century* (Chicago, 1938).

Packard, A. S.—*Lamarck* (London, 1901).

Pagel, W.—"William Harvey and the Purpose of Circulation," *Isis,* Vol. XLII (1951).

Patterson, L. D.—"Hooke's Gravitation Theory and its Influence on Newton," *Isis,* Vols. XL-XLI (1949-50).

Pledge, H. T.—*Science since 1500* (Science Museum Publication, London, 1939).

Randall, Jr., J. H.—"The Development of the Scientific Method in the School of Padua," *Journal of the History of Ideas,* Vol. I (1940).

Revue d'histoire des sciences, juillet-décembre 1951. Special issue on " 'L'Encylopédie' et le progès des sciences."

Rosen, E.—"The Commentariolus of Copernicus," *Osiris,* Vol. III (1937).

Three Copernican treatises (New York, 1939).

Sarton, G.—*The history of science and the new humanism* (Cambridge, Mass., 1937).

Schneer, C.—"The Rise of Historical Biology in the Seventeenth Century," *Isis,* Vol. XLV (1954).

Singer, C. J.—*The evolution of anatomy* (London, 1925).
A short history of medicine (Oxford, 1928).

Snow, A. J.—*Matter and gravity in Newton's physical philosophy* (London, 1926).

Stoner, G. B.—"The atomistic view of matter in the XVth, XVIth, and XVIIth centuries," *Isis,* Vol. X (1928).

Syfret, R. H.—"The Origins of the Royal Society," *Notes and Records of the Royal Society,* Vol. V, No. 2.

Taylor, F. Sherwood—*A short history of science* (London, 1939).

Thorndike, L.—*History of magic and experimental science,* Vols. I-VI (London, 1925).

Science and thought in the fifteenth century (New York, 1929).

Whitehead, A. N.—*Science and the modern world* (Cambridge, 1927).

Wightman, W. P. D.—*The Growth of Scientific Ideas* (Edinburgh, 1950).

Index